Be Your Own Home Renovation Contractor

Be Your Own Home Renovation Contractor

Save 30% without Lifting a Hammer

CARL HELDMANN

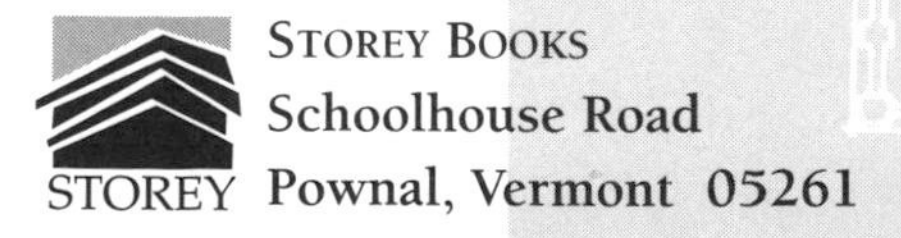
STOREY BOOKS
Schoolhouse Road
Pownal, Vermont 05261

The mission of Storey Communications is to serve our customers by publishing practical information that encourages personal independence in harmony with the environment.

Edited by Elizabeth McHale
Cover design by Meredith Maker
Photographs on pages 3, 4, 5, 7, 14, and 51 by Pauline Guntlow; on pages 50 (bottom) and 105 by Cindy McFarland; on page 50 (top) by E. Scolfield; on page 52 by John E. Stovel; and on pages 103 and 104 by T.J. Taft
Text design and production by Cindy McFarland and Erin Lincourt
Line drawings by John M. Knight
Indexed by Randl W. Ockey, Writeline Literary Services

Previously published in 1987 by Storey Publishing as *Manage Your Own Home Renovation: How to Save 30% without Lifting a Finger* © 1987 by Carl Heldmann; originally published in 1983 by Putnam Publishing Group as *Manage Your Own Home Renovation Project* © 1983 by Carl Heldmann.

Printed in the United States by R.R. Donnelley
10 9 8 7 6 5 4 3 2 1

Library of Congress Cataloging-in-Publication Data

Heldmann, Carl.
Be your own home renovation contractor : save 30% without lifting a hammer
p. cm.
Previously published as : Manage your own home renovation. c1987.
Includes index.
ISBN 1-58017-024-2 (alk. paper)
1. Dwellings—Remodeling—Amateurs' manuals. 2. Construction contracts — Amateurs' manuals. 3. Contractors—Selection and appointment —Amateurs' manuals. I. Heldmann, Carl. Manage your own home renovation. II. Title.
TH4815.H35 1998
643'.7—dc21 97-48318
CIP

Dedication

To my wife, Janie . . . who restores my faith in everything.

For their devotion and time,
I wish to thank my wife, Jane Prante Heldmann,
and also Cella Hunt Prante.

My very special "thank you" to Joyce F. Carpenter.

Contents

Part III Getting It Done

Appendixes

Introduction

There is probably nothing so satisfying as seeing an old house or structure restored to its original beauty. In renovating an old structure, you have the thrill of seeing the "before" and the "after." You also feel the accomplishment of preserving part of the past.

Whether you physically do the work, or you employ others to reach your goal, the feelings are the same. So are most of the savings, for there is probably no better way to save than to take another's junk and make it your treasure.

This book will show you how to obtain your treasure by restoring an old structure. It will show you how to do so without doing any of the physical work yourself and without possessing the technical knowledge of the various trades and skills involved in the renovation process. Do you know a lot about plumbing, electrical wiring, or heating? Neither do I, and I've been a general contractor for many years. What I do know, and will pass on to you, is how to find, hire, and schedule skilled tradesmen. I'll also show you how to purchase the materials needed for restoration.

I get paid well for what I know, and, therefore, you can have what a general contractor would normally earn in the form of salary, profit, and overhead. Since you won't have my overhead expenses, you'll actually be saving more than I would gain, as all your savings will be profit to you. The amount you will actually save (discussed in chapter 2) is considerable. These savings can be applied as your down payment, or to lessen the amount you'll need to borrow (discussed in chapter 7). It may make the difference between whether or not you can afford the project and/or qualify for a loan. I certainly hope it does make that difference and helps you succeed, for then this book will have accomplished its primary objective, and it, too, will succeed!

Part I

Making the First Decision

Chapter 1

Finding a Restorable Structure

Rehabilitation, restoration, remodeling, and renovation are words that mean very close to the same thing. Rehabilitation means to bring back, as does restoration. Remodel and renovate indicate changing something. I am going to use all four words interchangeably at times, so if you are a purist, please forgive me. The word I'll be using the most, however, is restoration. I call an old structure that can be recycled or made habitable an R.O.S., a restorable old structure.

Restorable Old Structures

First, what is an old structure (O.S.)? For the purpose of this book, it shall be any building whatsoever which, due to old age, neglect, or having been used for purposes other than habitation, is not suitable for habitation today. Or, if it is habitable, it is barely or not preferably so. Or, an O.S. can simply be a house that needs renovation. This means that our O.S. can be an inner-city, country, or even a suburban structure. Inner-city O.S.'s can be brownstones, row houses, old stores, apartments over stores, lofts, even factories, and bungalows. A country O.S. can be a barn, an old farmhouse, a carriage house, an old mill or anything else one happens to find when scouring the countryside — even a church. A suburban building can and most likely will be an older house, although there are many forms of old commercial structures in the suburbs.

△ **Before:** *Sometimes in house-hunting, "ugly ducklings" are the best buys.*

△ **After:** *Once renovated, these houses can become the pick of the litter.*

△ **Before:** *This cottage dates back to the 1920s — and so does its asbestos siding.*

△ **After:** *With new siding and accent colors for the trim, this "old structure" becomes up-to-date.*

Before: *The back of this cottage is in need of a face-lift, too.* ▷

◁ **After:** *The look of the back entrance is transformed simply by installing a 9-foot window and deck.*

After: *Lattice work hides the cinder-block foundation, cleaning up the area around the porch.* ▷

All R.O.S.'s will have one or all of the following, regardless of the state of disrepair of each:

- ▷ Exterior walls enclosing a space sufficient to inhabit, even if an addition is planned
- ▷ A roof, even if only the floor of a structure above
- ▷ A foundation, even if it belongs to a structure below

But how old is old? That is a rhetorical question, for any item can become old before its time. Literally, old refers to the later part of the life or term of existence of something. An O.S. could be relatively new in age, but, because of neglect or poor quality, near the later part of its life.

Don't overlook any possibilities by confining your search to aged structures only. Some O.S.'s that are a hundred years old can be in better condition than those twenty years old. Many suburban neighborhoods of houses built in the forties and fifties have badly deteriorated and should definitely be considered in your search for a restorable O.S.

That brings us to the question of what an R.O.S. really is. I define an R.O.S. as an O.S. that can be restored (rehabilitated, remodeled, or renovated) at a reasonable cost. Cost will be the primary determining factor as to whether or not an O.S. is, in fact, an R.O.S. It is not the only factor, for location, aesthetics, and personal needs and tastes are important, too. Obviously, if you have an unlimited budget, or for reasons of proximity or aesthetics want to restore an O.S. even if the cost becomes unfeasible, you should use other determining factors that I will discuss shortly.

Except in rare cases, the cost of the process of restoration, rehabilitation, or remodeling plus the cost of the O.S. should not exceed the cost of a comparable new structure. Therefore, an R.O.S. is an O.S. that can be restored at a cost that makes restoration feasible. We will discuss costs at length in chapter 5 and feasibility in chapter 4.

You may be surprised to discover what is restorable. A house totally burned to the ground can be restored less expensively than a new one can be built. That's an extreme example, of course, but the reasoning applies to O.S.'s in better condition. Obviously, if you start with *something*, you will have more than if you started with *nothing*. In the case of the burned-out house, you would be starting with a foundation, concrete slabs, driveways, and a cleared lot. If you have to pay full market value (new value) for this O.S. in such a state, it would behoove you to go elsewhere. But if you could purchase it quite cheaply, as you would most likely be able to, you'd be ahead of the game.

△ **Before:** *This house needs extensive renovation, but if done in stages, it is quite manageable.*

△ **After:** *The result is well worth the effort.*

Many factors will affect your decision to purchase and improve an R.O.S. and those factors will in turn increase or decrease your total projected costs. In many cases, if you are willing to live with the inadequacies of an unmodern structure, you can do the bare minimum to an O.S. and get by.

If you are willing to live with inadequate plumbing, heating, wiring, and no air conditioning, you can often find an inhabitable O.S. that, with a few minor repairs, a good cleaning, and new paint, is entirely livable. I call these minor repairs cosmetic improvement. You can, at a later date, restore parts or all of this O.S. as time and money permit.

Finding an O.S. for which cosmetic improvement will suffice, is more difficult than finding an R.O.S. in need of more costly repair. But they exist and can be found. Obviously, these O.S.'s are, or were, dwelling units to begin with, as opposed to nonresidential O.S.'s that require conversion. Some good examples of structures suitable for cosmetic improvement can be found among farmhouses, row houses and brownstones, or one-family houses, especially in old neighborhoods.

Why Buy an Old Structure?

The obvious reason is, of course, to save money. An R.O.S. presumably will cost less than a comparable new structure, especially if you are your own general contractor. There are other reasons for buying an O.S. that are quite valid, but these will vary in importance with each individual. Just as some people are fond of antiques and find in them great warmth, beauty, or quality, many see a beauty in O.S.'s that others don't. Their prime concern in restoring an O.S. is aesthetics; cost considerations are secondary. Some may be interested in an O.S. because of its proximity to their work, shopping, or friends. In most urban areas if proximity to the heart of the city is desired, an O.S. is one of few alternatives. Others may want just the opposite — the country is for them. Preservation of the past is a motivating concern to some. They will restore an O.S. to save it from demolition. Many feel that the housing developments built in the past thirty years leave much to be desired, lacking individuality, warmth, and charm.

The person on a limited budget who is interested in basic shelter and concerned with aesthetics and other considerations, may have to temper selection by the realities of budget. If budget doesn't have to be your primary consideration, these other factors can play increasingly more important roles. However the underlying theme of this book is how to save money, and I shall assume that saving money is the dominant reason for most people to attempt the restoration of an O.S.

How much you will save overall will be determined by how much you pay for the O.S. and how much you have to do to it. Buying right and wise restoration are the keys to saving and are fully covered in later chapters. Saving the general contractor's fee is the primary savings but knowing *how to buy right and how to reduce cost in the actual restoration process* can be equally rewarding financially (see chapter 6).

You can make a remodeling, rehabilitation, restoration, or renovation project as difficult or easy as you want. The main disadvantage in doing a project, as opposed to buying a new property — if indeed it can be called a disadvantage — is, if *you* permit the project to become too complicated, it can overwhelm you. I will show you the simple approach and encourage you to maintain the right frame of mind. You must bend and compromise when necessary and not become discouraged.

Another disadvantage exists if previous restoration in your selected area has driven up the price of an O.S. to a point where restoration is no longer feasible. This does happen and it is unfortunately becoming more prevalent as numbers of people undertake restoration projects. It usually occurs by neighborhood, though, not by city, so you merely have to extend your search to other areas. This, in fact, may be your first compromise.

At the other end of the spectrum, you may be the "pioneer," the first newcomer (or close to it) to a run-down neighborhood. This, on the plus side, always means a lower price for the O.S. — often ridiculously low — but poses two immediate problems that need to be faced. One is security. The other is the possibility that no one else will follow your example and the neighborhood will fall further into ruin. The problem of security is discussed later in this chapter. The second problem is extremely unlikely, for as the cost of new housing rises, O.S.'s everywhere will become more and more valuable. It is doubtful that any neighborhood will fail to be restored in the years to come. This too, is covered in more depth later in this chapter.

Another disadvantage is when there is too much work necessary to restore an O.S. This is not really a problem, for feasibility will be one of your first considerations (covered in chapter 3), and if there is too much to do, you can forget that particular O.S. and continue shopping.

So there really aren't many actual disadvantages in purchasing an R.O.S. but there are a lot of potential advantages — especially with you acting as your own general contractor.

Remember that even during a housing recession there is truly a buyer's market simply, and paradoxically, because *few* people are buying. When few people are buying, sellers become extremely apprehensive. When sellers become apprehensive, they do things they never would do if people were buying. They lower prices, help with financing, and agree to special arrangements. That's why it is a buyer's market and to your advantage to buy now, although this market condition may exist for a long time. You will overcome high interest rates and costs by managing your own renovation project. The buyer's market will help you buy your R.O.S. at the best price.

Use Your Imagination

Some people can look at something that is unfinished and envision precisely how it is going to look when finished. An artist, for example, knows this when the painting is just begun. I, as a builder, know precisely what a house is going to look like before I ever begin. In my mind's eye, I picture it sitting on a particular lot. I also can see a shabby wreck of a structure and picture it totally restored — even to the colors — all in my mind. I feel all of us have this ability to some extent. It is enhanced by experience, but all experience does is give us the confidence to use what was there all along — our imagination. If you allow yourself to try, you will be able to look at an R.O.S. and imagine what it will look like restored, even if you have never tried before. You must do so or you won't even be motivated to attempt the renovation and you won't have that thrill of seeing your goal, the finished renovation, in front of you.

In searching for an R.O.S., seeing one or two restored structures will help nourish your imagination. For example, if a particular section of a city is in the midst of a renewal and several homes, perhaps row houses or brownstones, have been restored to their original beauty, you can get a better feel for what to expect yours to look like. You also know that something good is happening to the neighborhood and your decision-making process is easier.

Some people are pioneers in spirit, have vivid imaginations, and don't care where they find an R.O.S. They are likely to have been among the first in a renewal neighborhood, or to have found an old mill out in the country and proceeded to renovate it without any qualms. Whether you choose to buy a deserted brownstone or a barn will ultimately depend upon a combination of personal taste and cost factors, but imagination is important in finding an R.O.S. for it allows us to visualize the exterior in its completed form. It isn't necessary to visualize how the interior will look, and quite honestly, some of them will look so bad you may not be able to envision a new interior. The inside design will be visualized and drawn out in plans by your architect or designer. Of course, some R.O.S.'s will be in good enough shape that no imagination will be necessary. Just don't get turned off by all of the filth and trash that you may find inside an R.O.S. or by holes in walls and floors, missing windows, ripped out plumbing and wiring, and so on. That is all relatively minor as you will soon realize. The best way to look at the inside of an R.O.S. is with your eyes partially closed, with just a fuzz of light filtering through your eyelids, and *then* use your imagination. If it is really bad, close your eyes completely!

Making an Informed Search

Obviously, the final decision to buy and restore an R.O.S. is yours, but it helps to have preliminary information when you are searching, or deciding whether or not a given structure in a particular area is a wise choice or if money is going to be available to you or if you can afford to do it. Advice on these matters is absolutely free and should be sought out and added to your knowledge of your own situation.

Real estate brokers can give you good input regarding a structure in a particular area. Be sure that their input isn't tempered with self-interest. (See chapter 6.)

Savings and loans can give you the same information as real estate brokers and it may be less biased. They can also discuss your money needs and their loan qualification procedure (see chapter 7). A visit to a savings and loan office would be one of my first steps in deciding to restore an R.O.S. Find a lender you feel comfortable with. Savings and loans are not all created equal and certainly not staffed equally. If you feel uncomfortable with one, try another. You'll have a rapport with one sooner or later. Don't be intimidated. Their business is to lend you money; that's how they make money.

Before you make a decision concerning an R.O.S., I strongly urge you to read this entire book thoroughly. Be sure you understand what the whole process entails from decision making to moving in. You really can't make a judicious decision, and you shouldn't attempt to, until you are as well informed as you can be! The books I recommend for further reading are listed on page 160. They are more or less aimed at the do-it-yourselfer, but they are good and will increase your knowledge of R.O.S.'s. Much of my own knowledge has come from books. But no single book is going to make you an expert. I'm still reading and still learning after ten years in the business. I learn because I am interested. I assume you are interested also because you are reading this book. So gather all the knowledge you can and decide what is best for you. I hope you arrive at a decision to renovate an R.O.S., for the purpose of this book is to help you do it and to save money on the project.

Prioritizing Your Search: Location, Cost, Design

Priorities need to be established in order to find an R.O.S. that is suitable for you. You will establish these priorities, but they will be based on where you want to live (location), how much you can and want to spend (cost), and what type of R.O.S. suits you (design). To help you establish these priorities, read this book, talk to other people, and consider your purely personal preferences.

Where to look is going to be based on whether you want to live in, near, or far away from the city. This then is the starting point and the first priority — location. Decide this point first and let the other two priorities, cost and design, follow in that order. There will, of course, be exceptions. This is the way I recommend; it is not a maxim. It is based on the premise that you will be able to find a suitable R.O.S. in both cost and design in your chosen location. This is not always true and when it's not, you may not be able to live in the location of your choice if you are to have an R.O.S. that is suitable in cost and design. Or you may "happen" across an R.O.S. that you fall in love with in a location that you never would have considered. By the way, falling in love with an R.O.S. *before you make an offer to buy* is very dangerous. You can also have an R.O.S. moved. House moving is expensive, but it is done every day. Professional house movers are listed in the Yellow Pages.

Cost, of course, will be determined by what you can afford to spend. I will give you all the guidelines; the rest is up to you. In chapter 5, we will go over a quick and easy method of estimating a project that you can use as you walk through an R.O.S. or while you sit in your car. It will give you a ball-park figure that should be accurate enough to tell you if the R.O.S. you are looking at is within your cost limitation. You must let your budget be your guide in the end — unless you have an unlimited budget. However, I have found that even extremely wealthy people have a budget and adhere to it.

Skipping ahead to design, I would no more recommend a particular style or period than I would recommend what color shirt you should wear. I will make recommendations here and throughout this book regarding interior space design, as this is often a matter of practicality rather than personal preference.

When your three priorities are established at least tentatively, you could then do what I would do — turn the search process over to one of my key individuals, the real estate broker. I would tell the broker what my priorities are and request that the search be confined by the constraints that those priorities impose.

Unless you have the time, and it might require more time than you think, I recommend you use a real estate broker. But be sure you explore a few other avenues that won't take too much time, before you employ a broker. These avenues include county agencies that may be involved in rehab housing or commercial lending institutions (banks and savings and loans). In larger cities you will more than likely find both. They generally involve themselves with inner city rehab housing and are often the source of good information about the location of very inexpensive R.O.S.'s.

Whether you prefer the city or country is a decision that only you can make. Such factors as proximity to work, shopping or transportation, the presence of noise or pollution, and your land desires,

△ **Before:** *This suburban home comes with a yard and two tall pine trees, protecting it from wind.*

△ **After:** *By adding a fourth window in the front for balance, wooden molding around the door and edges, and masonry to the exposed smoke stack, this home takes on a colonial feel.*

will help determine this on an individual basis. The price of the R.O.S. will greatly influence your decision. If you can't find one at a suitable price in the city, you may have to alter your priorities and look in the suburbs or country.

What to Look for in Buying a Restorable Old Structure

There are several items that are important when looking for an R.O.S. First, the structure. (This will be covered in depth in chapter 3.) The actual in-depth examination of an R.O.S. needs to be done by a pro, another of your key individuals, the inspection engineer. But there are a few very simple things that you can look for that will allow you to make a quick go/no-go decision.

▷ The neighborhood — What is it like?
▷ Zoning — What will the neighborhood look like in the future?
▷ Fire insurance rates — Are they high, low, astronomical?
▷ Property taxes — Will they rise rapidly or remain steady?
▷ Security — Is there loitering in neighborhood, windows barred?

The neighborhood can be very important so while you are examining your prospective R.O.S., take a look around. Would you like to live here? Better yet, would you, or could you, if you had to? Maybe it is in the midst of rehabilitation and you love it. At any rate, look — it is important. What's over there a block away — a trash dump? Who are your neighbors going to be, now and in the future? Will you like them? Will you fear them?

Some of your questions won't have an answer, at least not an immediate answer. You will need to talk to others. If there are people there who have already renovated an R.O.S., talk to them. They won't mind. In fact, they will be glad to see more potential rehabilitation coming into the neighborhood. It enhances their property value. Of course, that could prejudice their advice to you. Talk to lenders who may be pushing for rehabilitation in the area, talk to your real estate broker if you used one. If not, call a real estate agency that handles sales in the area. Most real estate brokers will be honest with you; their reputation is at stake.

In the country you can usually be less concerned about the neighborhood, but you still need to look around. Is that a pig farm down the road? Very odorous, indeed!

Zoning is supposed to mean that what you see around the neighborhood today, you will also see tomorrow. That's a rather flippant definition of a serious topic, but it is true. Zoning is important and yet some cities and locales don't have it. You (or your real estate broker — another reason to have one) will need to check to see if your locale has zoning and if it does, what it does for you. Zoning is supposed to keep residential neighborhoods residential, business neighborhoods business, and industrial neighborhoods industrial. But often there is spot zoning allowed which mixes business with residential. Sometimes that is good, often it is bad. In the country, zoning is just as important. Do you want a gas station built next to your restored farmhouse?

Sometimes other things, such as *deed restrictions*, do the same as zoning and these can protect you, but not for as wide a geographic area as zoning can. Your attorney and/or your real estate broker can advise you about zoning and deed restrictions. It is their field of expertise. That's what they are there for — use them!

Fire insurance is an item I hesitate to mention, for it is a rather negative one, but in good conscience I can't leave it out. Some areas are considered a worse fire risk than others, and this can seriously affect the amount of money one pays for fire insurance. The rate that one pays is partially determined by how long it takes the fire department to respond and the probability of fire in a particular neighborhood. Obviously, if you live twenty miles from the nearest volunteer fire department, the chances are slim your home could be saved in the event of fire (although most likely *you* would be saved by a smoke detector). The insurance company would then have a large claim to pay. Therefore, you will pay higher rates. The same would be true if you lived in a neighborhood with a high rate of arson. Many factors affect rates, and you will need to make a quick call to your insurance agent to see if there is a problem. Mostly likely there won't be, but better safe than sorry.

Property taxes are usually based on the property's value, so these taxes can be, and often are, escalated drastically as a neighborhood and/or a dwelling is improved. Some locales, in order to encourage rehabilitation, promise little or no increase in taxes. Some couldn't care less and tax away at the new value you have given (or will give) your R.O.S. Be sure to call your county or town tax agent, or have your real estate broker find out what your taxes will be now and what they will be when your project is finished. They could increase by over $1,000 per year, so be careful.

Security can be a problem. But this is the last discouraging part of the book, so cheer up! Some R.O.S.'s will be found in isolated areas, in or adjacent to run-down neighborhoods, or near commercial areas. This often, but not always, presents a security problem.

In the case of inner-city rehabilitation (urban renewal), there is often less crime as neighbors are more watchful due to a more acute awareness of potential crime. Most crime in this country consists of daytime robberies and no area — new or old, rich or poor — is immune. If I were concerned about a particular area I would talk to the police, neighbors, real estate brokers, and savings and loans which lend in that area. I also recommend considering the installation of a burglar alarm. There are many models to choose from, and they range from simple to sophisticated. Their costs range accordingly. You can do the installation yourself with components from an electronics store, or you can call a security alarm company listed in the Yellow Pages.

If the O.S. is suitable to you by these criteria you or your real estate broker need only quickly look for the very obvious defects that could prove costly. Until your inspection engineer does a thorough exam, you won't be able to determine actual restoration costs. But to keep you from needing to call the engineer more than once or twice, first look for these obvious defects:

- ▷ Serious fire damage
- ▷ Severe sagging of the roof
- ▷ Severe leaning of a wall or walls
- ▷ Crumbling or wide cracks in the foundation
- ▷ Serious water damage from roof or plumbing leaks

What is severe? How wide is wide? There is, unfortunately, no concrete answer (pardon the pun!). Let common sense prevail and remember, it is going to be examined by a pro. Your main concern will be to select an O.S. that is in a place you are willing to live and in good enough condition that your construction engineer won't think you are *non compos mentis* when he sees it.

When to Look

My wife and I were looking at an R.O.S. we were considering buying. Everything seemed fine — price, design, location, amount of restoration needed — everything. We saw it at 2:00 P.M. and fell dangerously in love with it. But the next morning I went by at 8:30 A.M. to have another glance and I was flabbergasted. What was a quiet tree-lined street at 2:00 P.M. the day before, was now a virtual freeway of rush hour commuters who had discovered a shortcut through from a busy thoroughfare. I checked again at 5:30 P.M and the same was true. We have small children and I'm not particularly fond of heavy traffic and noxious fumes especially in the morning, so we did not pursue our dream R.O.S. If traffic bothers you, check at different times of the day and on weekends. Also check for parked cars in the evening. You might have to park ten blocks away if you get home later than your neighbors. Neighborhoods change at different times of the day — keep it in mind.

Buy the Worst in the Best Neighborhood

It is probably more judicious to say "buy the cheapest in the best neighborhood." That way you allow for the fact that the worst isn't always the cheapest and vice versa.

Why would you want the cheapest? It is a real-estate fact of life that the lowest priced house in a neighborhood *appreciates* the fastest and is more assured of appreciation in the first place. One should always think of resale value of a home, and if you own the most expensive home in the neighborhood it is usually the hardest to sell. This happens for two reasons: (1) as the price gets higher, the market, or number of *qualified buyers*, gets smaller; (2) many people are aware of the fact that you shouldn't buy the most expensive house in the neighborhood. More and more people are realizing that they don't need large homes — at least not as large as they once would have considered. High energy costs and changing lifestyles that leave less time to care for a large home have precipitated this thinking. So stay toward the low- to middle-price range if possible. This also means that you will probably be purchasing a smaller R.O.S., making your restoration costs lower. As you will see, there is a direct proportion of size to cost.

Chapter 2

What It Means to Manage Your Own Renovation Project

How Much Will You Save?

The primary reason to be your own general contractor is to save money. You will find in reading this chapter that there are several ways in which you cut costs by managing your own home renovation. As your own general contractor you:

- ▷ Avoid paying a general contractor's markup
- ▷ Avoid paying retail price on certain materials
- ▷ Avoid paying a builder for unforeseen contingencies
- ▷ Are able to deduct sales tax paid on all materials from your income tax

By assuming the duties and responsibilities of a general contractor, you will save a dollar amount equal to what a general contractor would have made in the form of profit and overhead. A general contractor usually takes the cost of labor and materials and adds a 50% markup. This means a 33⅓% gross margin, or one-third of the total cost. For example: $500 cost of labor and materials X 50% markup = $250 for profit and overhead. So, $500 cost of materials and labor + $250 profit and overhead = $750 total paid by the client. As your own general contractor you would save $250; one-third of the total cost you would otherwise pay.

Markup and gross margin have always confused me as I am sure they will you. Don't worry about it. You'll save the same amount of dollars whether or not you know what it is called. Also, you will not have the normal overhead expenses that a professional general contractor would have, such as a truck, car, business phone, accountant, and licenses. Consequently, you can consider the entire amount of profit and overhead as savings to you.

If you were to hire a general contractor on a *fixed price contract* to handle the entire project without involving you, other than in selecting decorative items, more than a 40% gross margin would probably be figured into the total amount. Using a general contractor you would also be paying retail price for some of the materials used, since the general contractor makes a higher profit on retail items. So, by being your own general contractor you will also save money on some of the materials.

If you were to establish the cost of your project (based on the methods shown in chapter 5) to be $39,000, a general contractor would probably charge an additional $26,000 on a fixed-price contract basis. Actually it could even be more, as the general contractor would not pass onto you all of his materials discounts, as I just mentioned. That could be two or three thousand dollars more. So, on a fixed-price contract, that $39,000 job will cost you $65,000 to $70,000. (These savings are not taxable.)

On a fixed-price contract, the builder is the only one who knows the cost and he could add another 4% to 10%, unbeknownst to you, to cover unforeseen contingencies. This is quite justified as you'll understand after reading this book. At any rate, that $39,000 job could cost you up to $70,000.

Now we introduce an extremely important fact concerning this amount you will be saving — or paying a general contractor. For the sake of working with simpler numbers, let us assume you will be saving $10,000. If you were to pay that $10,000 to your general contractor, you would have to pay it with after-tax income. Assume that you are in a 30% tax bracket. That means you would have to earn about $15,000 before taxes in order to have $10,000 to pay someone else. Fees to general contractors are *not* tax deductible. So when considered in the light of after-tax dollars your savings are even greater. Your specific after-tax savings will, of course, vary with your individual income tax rate.

Another place you will save is deducting from your personal income tax any sales tax paid on materials you purchase. This amount will vary according to local sales tax and can be anywhere from a few hundred to a few thousand dollars. It is a deduction to which you would not be entitled if a general contractor manages your project and pays the bills.

You'll also save something besides money — peace of mind. You'll proceed at your own pace, and not at the schedule of someone else. You will also get exactly what *you* want done at a cost *you* decide upon and most importantly, *you* will control the quality of the work and materials.

All these savings require you to do a job — for which you are going to be paid quite well considering the hours you'll be involved. Let's see what it is you have to do, what you have to know, and how much of your time is required.

What Skills You Need

Usually I injure myself when I attempt a do-it-yourself project around the house. I have grandiose ideas for projects but when I do get around to doing them, the result is usually poor. I am like many general contractors. We aren't tradespeople or very handy do-it-yourselfers. We are professional managers of people, time and money. Those are the required skills for managing your own renovation project. Consider them as just one skill — a manager's skill. The physical skills that are required, whether they be electrical, plumbing, architectural, or legal are possessed by the professionals we hire, our subcontractors. They are the masters of their trades. (I'll help you be sure of that. We will discuss subcontractors at length in chapter 8.) Your job is to hire and manage those people. Don't let that scare you. It's quite easy. They know their jobs, I'll help you know yours. And to answer that question that continually pops up

— *you do not need a license to contract (hire) people when it is your own home or land.* Period! You do need a license, usually, if you perform the same function for someone else on property that is not yours. But, I repeat, when it is your home — and it will be when you buy it — you do not need a general contractor's license. You will need several permits, but you can easily get those, and I show you how in chapter 10.

Do you have this manager's skill? You probably do and you probably use it daily. Think of the countless times during a day when you have to deal with people — at the supermarket, on the phone, at the store. Most are people trying to sell you something or perform a service — suppliers and subcontractors. Your entire day is spent managing time and people and day-to-day expenses. All of this is good preparation for a general contractor. Think of your own job or career. Can you see these management skills appearing among your capabilities from time to time, if not constantly? They are not too difficult to master, but you don't even have to master them to renovate one house. You just have to be aware of what the skills are. I have seen general contractors who can't balance a checkbook. I even know a few general contractors who don't know how to manage people, nor do they care. As for time, remember this old cliché, "If *you* take care of everything else, time will take care of itself."

How Much of Your Time Is Needed?

While we're talking about time, let's see just how much of your time will be involved. A few months ago, my neighbor, a housewife, told me she wanted to handle her own remodeling job. She felt comfortable dealing with subcontractors, as she had hired a plumber a few times to fix serious maladies and had hired other workmen occasionally. And she always had charge of household expenses. She even did the family tax returns. But she felt that she was so involved with volunteer work, carpooling, and other things that she wouldn't be able to spend enough time on the remodeling project. I asked her how much time she thought she would need to spend. She guessed she would have to be there — on the job — most of the day, every day. I will tell you what I told her. Most general contractors have several jobs going at once — they couldn't possibly spend more than an hour or two a day at any one job. Some days they don't visit any job site; there simply isn't any need and they have other things to do. You will not need to spend more than an hour or two a day actually visiting the job site, and then not every day.

Most of your time will be spent on the phone calling subcontractors and this is done in the early morning or in the evening. When we discuss subcontractors in chapter 8, you'll see that you don't have to be there to watch them work.

The need to physically check or inspect work in progress is almost nil and completed work can be inspected early in the morning, on lunch hours, in the evening, or on weekends. If you are gainfully employed, there is absolutely no way being your own general contractor will interfere with your job — unless you allow it to. *Don't.* Your investment of time *will* be minimal. Most of your job will be completed before anyone starts work on the project. Your organizing and planning will be 80 percent finished before renovation starts and I assure you, that's the main component of managing your own renovation project.

Begin by training yourself to look at parts of the project and not the whole project. This will bolster your confidence. Even today, I can become discouraged if I look at the magnitude of an entire project. But I know through experience that even the largest job is merely many little jobs put together, and that none of the little jobs is difficult by itself. If you look at it this way, and make sure all those little jobs get done — by others — you will succeed. You really can't fail. It is like driving a car from New York to California. The thought of driving that car turns me off. But if I look at it this way: Allowing seven days, driving three to four hundred miles per day; taking time for a little sightseeing and fun; looking back each day with satisfaction at how far I've gone, the trip takes on a whole new perspective, one that I would enjoy.

The "unknowns" that cause a very natural fear, will disappear. You will either learn those unknown things or find that they are simply unimportant. There are thousands of things I and my fellow general contractors don't know today. I don't particularly care what some of the technical terms used in certain trades are. It used to bother me that I didn't know, now it doesn't. If I haven't learned them, they are not important to me.

When you have finished reading this book, you will have sufficient knowledge to act as your own general contractor. Read it and reread it! There will be no reason you cannot successfully manage a renovation project.

Key People and Other Helpers

Whether you employ or act as your own general contractor, there will be a few key people whose services will be invaluable. I would not attempt a project without most of them involved at one time or another.

Throughout the renovation process you must rely on the expertise of the professionals you hire, as all general contractors do unless they happen to be, by profession, one of these key people as well as a general contractor. In a renovation project, as in most circumstances, penny-wise is dollar-foolish. Trying to get by without one of your key people in order to save a few dollars could prove to be costly. You won't have many, if any, surprises if you use your key people to their fullest. Surprises mean unforeseen expenses due to unforeseen problems. Enough said! Here are your key people.

- ▷ Inspection engineer
- ▷ Architect or designer
- ▷ Real estate broker
- ▷ Real estate attorney
- ▷ Experienced carpenter

Inspection Engineer. In your Yellow Pages, or through references from a real estate broker, home builder's association, savings and loan, or professional general contractor, you can find and hire an inspection engineer. This person is usually someone with an engineering degree and/or considerable experience. The engineer will inspect an older structure for defects, advise you of the corrective measures needed, and be able to roughly estimate the cost of repairs. (We will discuss more fully in chapter 3 exactly what these engineers inspect.) This key individual's services will be needed very early in your project. In fact, based on a negative report by an inspection engineer you may very well not have a project. Of all your key people, this one is, in my estimation, the most important. I wouldn't dream of buying an O.S. without a written report from an inspection engineer. Most guarantee their report and carry insurance that guarantees their guarantee. I know how to inspect, but I wouldn't dream of guaranteeing my own report. Don't be misled that you or a "knowledgeable" friend can inspect an O.S. No matter how many books you read on the subject, get a professional and get a guarantee. If you can't find an inspection engineer in your town, see chapter 3.

Architect or Designer. This is the second most important key person to come into your project. Just as you wouldn't drive from New York to California without a road map, you cannot I repeat cannot, successfully restore an O.S. without an architect or designer. His or her advice will guide you on your journey better than a road map would on a trip. The cost of hiring this key individual is more than compensated for in reducing wasteful and costly mistakes or procedures. The architect or designer will evaluate structural change considerations, make suggestions, integrate design with building integrity, draw blueprints and, if necessary and at an additional cost, oversee the entire project. (For more on this important individual see chapter 3.)

Real Estate Broker. You can conceivably do without a real estate broker if circumstances permit, but he or she is also an individual who can prove to be invaluable. In chapter 6 we'll discuss what a real estate broker does in the purchase of an O.S., but for our purposes a real estate broker is probably most important in the searching process. It is their job to find properties and many specialize in O.S.'s. A good real estate broker can save you more than his or her commission. At the very least they will save you an incredible amount of time looking for an O.S., as we'll see in chapter 6.

Real Estate Attorney. I am told that most people have never used an attorney and therefore are apprehensive at the thought of having to deal with one. I deal with them very often and I hope I can allay those fears. A real estate attorney will be quite important to you. He or she will handle the legal aspects of your purchase of an O.S. In any real-estate transaction it is advisable to use an attorney to handle the *closing*. To not use one is not only fool-hardy, it is economically dangerous, especially when dealing with an O.S., which may have had many owners or other complicating legal factors. (For more on attorneys see chapter 8.)

Experienced Carpenter. My knowledge of carpentry is very limited. I will assume yours is less. It is not the function of any general contractor to train a carpenter in methods or workmanship. The carpenter's work precedes all the other trades (except for laying the foundations), and this work is the most crucial. Experience has no equal in carpentry. The little tricks of the trade and knowledge of this key person are invaluable. In the long run they'll save you more time and money than if you hired a less experienced, and perhaps less expensive, carpenter. In chapter 8 I'll show you how to find one and how he will help you in other ways.

Other Professionals. The other professionals who will potentially help you are any of your subcontractors or suppliers, as we'll see in chapters 8 and 9, as well as county and town services, savings and loans, and even general contractors. They will all give advice when asked (and often when not). They are used to being asked for advice. Their suggestions are usually based on experience and they won't think less of you for asking. In fact, they'll think more of you, and best of all, it's free!

Yes, You Can Get Financing

In new homes and retail residential sales of previously owned homes, financing a sale has become the scourge of buyer and seller alike — not to mention the real estate broker.

High building costs, as well as higher interest rates, have raised the cost of the average house out of the reach of many first time home buyers — ones with no *equity* from a previous home. They can't qualify for a loan even if they earn good salaries. You are lucky. In considering an R.O.S., you are easing the two major factors that contribute to the housing dilemma. The first major factor is high cost. By buying an R.O.S. properly, as we will discuss in chapter 6, you will be buying well below market value. Because it is old and needs work, you have a definite advantage in controlling price. In some cases, where local government is involved, you may be buying an R.O.S. for pennies.

The second major factor, high interest rates, will be less problematic because you will be borrowing far less than if you were to buy an already restored house. After you buy it (for less), you are going to act as your own general contractor and save even more. This will lower your monthly loan payments, thus lowering the qualification requirements (discussed in chapter 7) of your savings and loan. Another benefit will be that based on what the market value of your R.O.S. will be when renovation is completed, you will have a considerable amount of equity — depending on how much you spend on restoration — even though you received this equity with little or no cash of your own involved. When the ratio of equity to loan increases on the equity side, savings and loans give lower interest rates on your loan. They do this because their risk goes down as your equity goes up. That means you will (or should) qualify for a lower interest loan than those who buy at retail and have little equity in their purchase.

Another point in your favor is that most savings and loans and commercial banks have recently taken a new view on restoration and are encouraging it, often with low-interest loans. This varies considerably with each city and town, but is well worth looking into as you'll see when we discuss financing in chapter 7. You will be able to arrange your financing to cover both the purchase of the R.O.S. and the amount needed for restoration. It is very simple when you are your own general contractor. What you save by managing your own renovation project will be your down payment. We'll also see in chapter 7 how to overcome any reluctance that a lender may show because you are acting as your own general contractor. You see, you will be able to get financing, and less expensively than you thought.

If You Do Use a General Contractor

If either time (you and/or your spouse are true workaholics) or fear (for shame! — reread the book) prevents you from managing your own renovation project, you can still save money. You have two alternatives. They both use a general contractor, thus relieving you of that responsibility. They do cost you money, but they reduce the general contractor's risk and he or she can and will charge you less.

Manager's Construction Contract. Under this arrangement, the general contractor is responsible for your subcontractors and will find, hire, schedule, and inspect their work when necessary. He will also pay fees, get required permits, arrange for any temporary utilities, and review your bills. You will be responsible for paying all bills including those for subcontractors. Very simple, yet you have relieved the general contractor of some big headaches — financial responsibility, cash flow, and bookkeeping. For that I normally charge half my normal profit and overhead. Your time involved under this arrangement should be no more than an hour or two per week — the time it takes to go over bills and write checks. (I will discuss paying subcontractors and suppliers in detail in subsequent chapters.)

An example of a manager's construction contract is in appendix A. You and your attorney can modify it to fit your particular needs. Remember, when you increase the general contractor's responsibilities, the cost of using a general contractor goes up. The reverse is true in reducing your cost.

Fixed Fee Contract. This contract is not to be confused with the fixed-price contract that I used earlier in this chapter as an example for determining how much you'll save by managing your own renovation project. A fixed fee contract is, as the name implies, a contract between two or more parties where only the fee for the general contractor is predetermined and fixed at a set amount. His fee is not based on percentage of costs (that's called a cost plus or percentage contract). In a fixed fee contract, the fee is not tied to the actual cost of labor and materials involved in the project. I stress this, as this type of contract lessens the general contractor's risk, and thus the fee, by eliminating the possibility he will lose money by estimating incorrectly. The general contractor will prepare an estimate, but it is not a guaranteed amount. If he is to guarantee that amount, or an amount close to it, he will charge more as a fee. He has to protect himself against those surprises I mentioned earlier.

When you demand and get price guarantees from general contractors, especially when remodeling, you are either going to pay more in the long run; not get what you think you are getting or what you want; or end up with a soon-to-be bankrupt general contractor, or at least a very unhappy general contractor, because he is losing his expected profit and overhead. He may cut important corners, leave things out, substitute materials, drag your project out too long while he looks for more profitable ones, or disappear altogether.

If, however, all you want is guidance and management ability, a general contractor will give it at a very reasonable fixed fee. This fee should approximate a percentage of the estimated cost without guaranteeing the estimated cost. Figuring 30 to 40 percent for profit and overhead is a reasonable amount and substantially less than the 50 percent markup you'd pay to have him handle the entire job. A general contractor will charge a higher percentage for a smaller dollar job, as the time spent is approximately the same as if it were a larger job. Sorry, but that's a fact of life and the contractor must earn a living too.

If you do use a general contractor, I recommend that you still read this book thoroughly so that you will be aware of what it is the general contractor will be doing throughout the project and what order of events will take place. It will add to your peace of mind. You will also have important information on planning, cost estimating, purchasing the R.O.S., and financing it.

PART II

Planning

Chapter 3

Is the Structure Restorable?

Several people will need to be involved in assessing the feasibility and extent of your renovation. An architect/designer, inspection engineer, loan officer, and carpenter all will be helpful in your decision-making process.

Architect/Designer

An architect/designer is more important in restoration than in new construction. The reason is, in new construction there are thousands of ready-made *building plans* and specifications available. In custom designing, of course, the need for an architect/designer is obvious, and in restoring an O.S., there are a few particulars that warrant an architect/designer. Of course if you are only going to paint and move in, you will not need an architect/designer. But once you decide to undertake major remodeling, restoration, or rehabilitation, you need a plan. This plan can be very simple or very elaborate. It will always consist of a master plan showing what is going to be done to the structure item by item, room by room. The master plan includes *blueprints* that show what the dimensions of each finished room will look like, and specifications that describe the materials to be used. Based on cost considerations, the master plan will determine what the blueprints and specifications will be. Theoretically, you could do the master plan yourself, but without spending too much money you can also get some expert advice from an architect/designer. The expertise and experience of an architect/designer can save you more money than it will cost, as the two of you determine together what to do to your R.O.S. You will both go through each room and decide what, if anything other than cosmetic improvements, to do to it. If walls are to be moved or eliminated, you will need the architect/designer's advice as to which can be moved without adding structural support. If additional structural support is necessary, either as a result of changes or because of an existing bad condition, he is the person who will design such supports. Other structural defects can be remedied on the architect/designer's advice, based on both the inspection engineer's report and his own evaluation. He will obviously have to physically inspect the R.O.S., too. You'll decide, based on this opinion, whether you can live with prevailing structural, plumbing, electrical or energy conditions, or if you should change them. His advice will guide you concerning the inclusion of such new technologies and modernizations as solar adaptations, saunas, and hot tubs.

The exterior of your R.O.S. will need careful evaluation for proper restoration based on aesthetics, structural preservation, and cost. Here again, an architect or designer's advice can be time- and money-saving. Perhaps you thought you had to do more than you really needed. Perhaps he will advise you not to spend $10,000 for new siding; all you need is a few rotten boards replaced, a sanding or sandblasting, and a paint job — for about half what you planned to spend. An extreme example, but viable, nonetheless.

Aesthetics are a very important consideration in an R.O.S. There needs to be expert guidance in determining what the finished project will look like. Without it you could end up spending a lot of money unnecessarily and have a terrible looking *restored* R.O.S. — called "remuddling." I have seen them and my first thought is "what a waste of time and money." This is not to say that you have no input about what your R.O.S. will look like. You will determine that. You simply need expert advice. You would gain limited expertise from reading books and journals on old house restoration, and it wouldn't hurt to do so, even if you employ an architect/designer.

But there is no substitute for the give-and-take in a conversation with an expert who can answer your questions and help you make decisions. Such invaluable suggestions as adding a dormer to make use of an attic, can come from your architect/designer, and give you inexpensive additional space.

One finds an architect or a designer either in the Yellow Pages or through referrals from friends, real estate brokers or lenders. A phone call will tell you a lot. I feel a friendly rapport is essential for working with this particular subcontractor. Did I say subcontractor? Yes, I did, because that's exactly what any person involved in your project is — a subcontractor. They are there at your request and are being paid with your money (even if it is borrowed) and you are employing them by contractual agreement. If you keep this in mind when you are talking to an architect, attorney, or any other of your subcontractors (subs), you will maintain the professional relationship necessary. This is not to say you should act haughty just because you are paying them. It simply means you are not to be intimidated by any sub or supplier — be it a professional person, carpenter, or savings and loan officer (I even consider them subs!). As I said earlier, friendly rapport is necessary but at the same time it should remain on a professional level with you in command.

Architects generally work for a fee based on a percentage of the project's cost or by the hour. With a percentage of cost, you run the risk of having the cost deliberately run higher than it need be. The percentage charged will vary depending on whether you want the architect to supervise construction or not. Obviously, it is going to cost you money for full supervision, but you don't really need it. Architects generally provide this service as a liaison between owner and general contractor. Since you are both, you don't need a liaison. You could still have the architect supervise if you feel inadequate or have him supervise on a part-time, as-needed basis. Otherwise, the architect's job will be finished prior to construction and will consist of only design, consultation, and blueprints and specifications. For this the fee would range between five and ten percent of project cost.

A home designer is less expensive but not necessarily less knowledgeable than an architect. He may work on the same fees/services plans described above, but designers will often work for a fee based on the square footage of the R.O.S. This fee usually includes only the design, consultation, blueprints and specifications. He will also supervise when needed on an hourly or percentage-of-cost basis, usually at a lower cost than an architect.

Again, proper advice and planning is extremely important and will save you money in the long run. Proper advice and planning can also show you ways to postpone certain items in order to save money initially. I'll discuss this in more depth later in this chapter.

Inspection Engineer

Prior to finalizing any arrangements with the architect/designer you will need the services of your inspection engineer which I described in chapter 2. The inspection engineer will provide the information that will enable you to decide whether or not to purchase your R.O.S. Conceivably, he and your architect/designer could inspect the R.O.S. together, but that is unlikely and an unnecessary expense. An inspection engineer or firm will, for a fee, inspect the following items and, most importantly, guarantee in writing the condition of each. Some inspections are even backed by an insurance policy and I would gravitate toward those who have such insurance. An inspection engineer should also be able to give you an approximate cost of repair for each item, although not an exact cost. (We will determine more accurate costs in chapter 5.) This approximate cost will give you a quick estimate that will tell you if you can afford the project. Most inspection engineers do not do repair work for the obvious reason that they would be inviting conflict of interest. Those who sometimes do will spell out explicitly beforehand that the inspection engineer will not be given any of the repair work on this project. You will get a more impartial and objective inspection that way. You will employ this person only when *you* have found an O.S. you believe to be restorable. You may go through this process more than once, I'm sorry to say, but it is better to spend extra money in the beginning than to get caught with a very expensive surprise later.

The items the inspection engineer will check and report on are:

- ▷ Foundation — for cracks, settling, water problems, crumbling, heaving
- ▷ First floor framing *sill* and *plates* and joists — for rotting, insect damage (termites), cracks, settling
- ▷ Exterior walls, siding, doors and windows, *soffit* and *fascia boards* — for rotting, cracks in veneers such as brick and stone
- ▷ Roof — for sags, rotting members, life expectancy of shingles;
- ▷ Insulation (a relatively new inspection) — to see what insulation there is if any, advise as to needs
- ▷ Fireplace and chimneys
- ▷ Water drainage on lot — especially if water damage has been found in basement or *crawl space*
- ▷ Plumbing — for water pressure, life expectancy of plumbing pipes, condition of fixtures (lavatory, closets, tubs, showers)
- ▷ Electrical — for type and condition of wiring, adequacy of service (power) to structure for current and future needs, condition and safety of electrical devices (receptacles and switches) and fixtures, all *panels* and their *fuses* and *circuit breakers*
- ▷ Mechanical — or condition and adequacy of any and all heating and air conditioning equipment and its life expectancy
- ▷ Appliances — (stoves, etc.) for condition of any and all
- ▷ Well and septic tank, if applicable — for recovery of well (water pressure over long flow period), visible problems with septic tank (standing water in leech or drain field, noticeable odors)

The foundation inspection can be separated from the rest and referred to as a structural inspection. Often that is available by itself at a lower cost, if that is all you need. I recommend the entire inspection if you are going to buy. If you aren't sure due to extremely poor condition of the O.S., then opt for the structural portion of the inspection only. If it passes, or passes for your purposes, then you can proceed with the rest of the inspection. If at all possible, accompany the engineer during the inspection. You can benefit by getting to know the O.S. better, and you'll pick up bits and pieces of information to help you if this one fails to pass muster and you have to continue searching. Don't worry if you can't accompany the inspector.

Testing for Environmentally Hazardous Materials

As part of a complete inspection program of an older home, tests should be performed to determine the presence of materials which could be hazardous to your family's health. Most dangerous are asbestos, lead and radon. Your State Department of Environmental Protection can be a good source of information for testing resources.

These tests should be the seller's responsibility, but may not have been done prior to the house going on the market.

Asbestos was used for its fireproofing and insulating properties in many houses built between 1920 and 1960, and most often can be found in pipe insulation, siding shingles and 9 inch square flooring tiles. If it's in good condition and it does not pose a health hazard, no laws or regulations require that it be removed. However, building owners are required to keep asbestos in good repair. If a demolition/renovation activity could cause damage to asbestos-containing material, then it is required that the asbestos be removed prior to the activity.

Removal and proper disposal is an expensive process and is required to be performed by licensed professionals. Asbestos siding and flooring can often be covered with new materials, but care must be taken that the asbestos is not broken and is contained under the new covering.

Lead. Almost all houses built before 1960 used lead based paint (which is now banned), and lead can also turn up in the water system where lead solder was used at pipe joints or from lead pipes in the water mains. Lead poisoning can be very debilitating, especially for children (those who like to chew on things are at very high risk).

There are a number of lead paint test kits available on the market for do-it-yourselfers, but since lead can actually be released into the environment by a sloppily done test, this is best left to a licensed professional who can also recommend remediation procedures if lead is found. Such procedures include encapsulation (covering with special paint), which is the easiest, to removal which is the most expensive. The safest method for small children is to remove all lead paint to a point 36 inches off the floor.

Homes built before 1930 are most likely to have lead plumbing. Lead in the water system can be detected during the water test, which should be done to detect other minerals and pollutants as well. The water to be tested should have been standing in the pipes for at least twelve hours before being drawn from the tap. A certified testing lab can be found in the Yellow Pages.

Radon is a naturally occurring radioactive gas that emanates from the soil and is the second leading cause of lung cancer in the United States. Nearly 1 out of 15 homes in this country have elevated radon levels, and it makes no difference if it's old or new, drafty or well-sealed or with or without a basement.

Radon can only be detected by testing, and testing devices can be obtained at retail stores. Since you are involved in a real estate transaction, you may opt for the short term testing method that requires a minimum of 48 hours for accurate results. Testing should be done at the lowest level of the home which will be used for occupancy. Make sure the device is certified by the U.S. Environmental Protection Agency (EPA), which also can provide you with more information on reducing radon levels in the home. Radon reduction can be expensive or not, depending on the design of the home. Radon Reduction Contractors are certified by the EPA or state agencies.

Compared to radon entering the home through the soil, radon entering the home through water in most cases will be a small source of risk. It is usually not a concern in public water supplies, but has been found in well water. If elevated levels have been found in the short term air test mentioned above, and the water comes from a well, have the water tested.

Advice from Key People

All your key people can offer advice and have specific areas of expertise that enable them to answer any questions that might arise in your decision-making process. By the time you have reached the point of determining if an O.S. is restorable, you should have lined up all of your key people and have talked to them about your project, even about this particular R.O.S. A lender, such as a savings and loan officer, can offer advice based on his experience. This advice can pertain to the worth of the neighborhood, how much you should spend, how much value added is too much, and so on. The only one you may not have lined up is the carpenter, but if you have, his advice can be beneficial, giving you a more accurate idea of cost pertaining to major woodworking structural changes or repair. The carpenter's time with you in consulting may or may not be free. That will depend on the individual and the scope of the prospective job.

All advice should be treated as such. It is an opinion as to what should be done. Some advice may be bad. The final decision on what will be done is yours and yours alone, but the more expert opinions you have on any matter the easier your decision-making will be.

Advice and opinions should not be confused with necessary technical decisions that must be made initially. No matter how many books you read on the subject or no matter how many expert friends you have, leave technical decisions to the professionals. A decision regarding the structural integrity of an O.S. and what needs to be done to rectify any problem is a decision that can be made only by a professional.

Partial Renovation

There is no "law of restoration" that says an O.S. has to be totally redone. It is often economically beneficial to do only those things that will make the structure habitable. This definition will vary for each individual as some can live without things that others can't and some can live with some things that others can't. Some things can be fixed later — years later. You can do one room at a time, or parts of a room at a time. One important criteria must be met, however; the structure must be safe in all ways — structurally, electrically, and so on. Some locales have habitation standards for dwellings, but some don't. We are considering so many O.S.'s, however, that the possible degrees of habitation are quite numerous. There is no set answer to what you should, shouldn't, could, or couldn't do in lieu of total rehabilitation or restoration. What is important to you will prevail, governed by your budget and common sense (and your inspection engineer's report). The need for a master plan is as great, if not greater, under these circumstances.Check with your local building inspector about how much of the total structure must be brought up to current code if only part of the structure is being renovated.

My personal preferences as to comfort would lead me to do as much to the kitchen and bath(s) as I could afford. Cleaning, painting and other cosmetics would then suffice. If the electrical wiring needed replacing, it can be done without removing the plaster walls. In fact, you probably can get by without replacing all the wiring. There are ways to do both and we'll cover them in chapter 11. Plumbing can be replaced without too much wall repair. Maybe the pipes will last another two years or so. The general idea is to do what is necessary *now* and put off other renovation. Of course, if you are starting with a barn . . .

Another point to keep in mind when considering partial remodeling is the total value of the end product. I have stated before that you don't want to make the house too valuable for the neighborhood. Partial restoration is a way to control your total cost and keep it from adding too much value too quickly.

What do you do when you find that your O.S. is an R.O.S., but it is going to cost too much or you feel it will be too expensive for the neighborhood? Well, if partial renovation can't or won't fit your particular situation, you will have to continue your search for an R.O.S. You must remain logical and unemotional (don't fall in love with an O.S.) in your decision-making process.

Chapter 4

Is the Renovation Feasible?

A **list of possible costs** that you may encounter in the renovation of your R.O.S. is given in chapter 5. Some of those costs will be applicable to your R.O.S., some won't. Based on your inspection report, advice from your other professionals, and your own common sense, judgment, and budget, you can determine which costs will be immediate, which can be eliminated, and which put off until later. In order to do this you need a priority list, a budget, building plans, determination of space requirements, a master plan, plans for additions, a completion timetable, and living arrangements during the project. Let us cover each of these components of feasibility.

Priority List

This is a method I use to place priorities on items pertaining to the project. I try to be objective, but personal preference will influence the list also. Beginning with the inspection engineer's report, I list all the items covered and the report concerning each. If the foundation is basically sound, but there is some settling causing floors to sag above, I need to determine if I want to fix this now or later. It can be done at either time. But if the foundation is crumbling and will probably fall or cave in soon, I would put that repair at the head of my priority list. This list is used to determine whether or not you go ahead with any costs, not just those pertaining to the structure. Use it to determine if you want an architect or designer, central air conditioning or window units, to purchase a new furnace or patch the old, to put in a driveway or not, to add more landscaping, and so forth. Clearly, that professional inspection engineer's report is crucial.

I mentioned that such problems as a sagging floor can be corrected later. That's true as long as the sag isn't too great. The inspection engineer should indicate this in his report. If it is too great, it will put unusual vertical and horizontal stress on the *framing* of the structure. Correction is not expensive.

SAMPLE PRIORITY LIST

	TOP PRIORITY (MUST SPEND)	LOW PRIORITY (OPTIONAL)	DON'T NEED (OK AS IS)
Inspection fees	✓		
Design assistance		✓	
Loan closing costs	✓		
Construction or short-term loan interest	✓		
Fire insurance	✓		
Temporary utilities	✓		
Cleaning the R.O.S.		✓	
Plans and specifications	✓		
Permits and fees	✓		
Clearing, grading, and excavation for additions			✓
Footings			✓
Foundations			✓
Waterproofing			✓
Tearing out			
▷ exterior walls			✓
▷ interior walls	✓		
▷ roof-shingles	✓		
▷ roof rafters			✓
▷ plaster, drywall			✓ (patching)

	TOP PRIORITY (MUST SPEND)	LOW PRIORITY (OPTIONAL)	DON'T NEED (OK AS IS)
▷ furnace, radiators	✓		
▷ plumbing	✓		
▷ electrical wiring, etc.	✓		
▷ total gutting			✓
▷ debris disposal	✓		
New framing — lumber, materials, and labor ▷ walls	✓ (2 walls)		
▷ floors			✓
▷ ceilings			✓
▷ roof			✓
▷ additions		✓	
Steel			✓
Windows	✓		
Exterior doors			✓ (refinish)
Roofing — labor and materials	✓		
Concrete flatwork			✓
Exterior trim — labor and materials	✓ (fascia)		
Exterior siding or veneers			✓
Plumbing — upfitting, per fixture	✓		
Plumbing — new, per room	✓		
Heating and A/C	✓		
Electrical — upfitting	✓		
Electrical — new wiring and service	✓		
Insulation — walls, ceilings, and floors	✓		
Drywall	✓ (patching)		
Plaster	✓ (patching)		

	TOP PRIORITY (MUST SPEND)	LOW PRIORITY (OPTIONAL)	DON'T NEED (OK AS IS)
Water and sewer main repair			✓
Well and septic system			✓
Cabinets, bath vanities	✓		
Interior trim materials			✓
Interior doors			✓
Interior trim labor			✓
Painting, interior	✓		
Painting, exterior	✓		
Appliances (built in)	✓		
Light fixtures		✓	
Floor coverings	✓		
Drives, walks, patios			✓
Cleaning		✓	
Gutters, screens, miscellaneous	✓		
Wallpaper		✓	
Hardware and bath accessories		✓	
Landscaping			✓
Miscellaneous allowance			✓

Special Notes: ____

____

____

____

____

____

____

____

____

____

____

____

____

What else can be lived with in order to put off the cost for repair or renovation until later? Other than what your inspection engineer says is important pertaining to the structure, you can live with just about anything that you are willing to put up with! Keep in mind when deciding what to save to do later that it will then cost you a little more to do. This will be due to inflation and the fact that small jobs cost comparatively more than big jobs. You will save overall by getting as much done at once as you can. Also keep in mind the inconvenience you may suffer from having your life disrupted by remodeling after you move in.

Ideally, if your budget permits, I recommend getting all the major jobs done now and pick and sort through the minor ones to see what to put off. The major jobs would be any that involve tearing the old out and replacing with new. This includes plumbing, electrical wiring, mechanical work, foundation, roof, chimneys, plaster, rotten framing, siding, porches, and any appliances that are either just about at the end of their life expectancies or aren't there at all. You don't have to completely remodel a kitchen in order to make it more convenient. A new stove or dishwasher and a coat of paint will do wonders.

If you have an R.O.S. that has nothing as far as kitchen or baths you'll have less to eliminate from your list. But you could eliminate decorative items such as wall paper, light fixtures, and landscaping.

If your inspection engineer says that the roofing is good for another two to five years, then wait if need be. It may not look so good, but so what?

Budget

All the decisions on your priority list are predicated mainly by the budget that you must prepare *before you even start looking for an R.O.S.* The total cost of the R.O.S. when finished shouldn't exceed what you have predetermined you can afford. You should be cautious in this stage, for it is easy to forget that the total or finished cost of the R.O.S. you are looking at can be considerably higher than what it is selling for now. You can't purchase an R.O.S. unobjectively and then find out once you have purchased it that the renovation will exceed your budget. This sometimes happens, but it won't if you calculate properly. If you determine how much you can afford to borrow (based on our discussion in chapter 7 on financing), and you know how much cash you have or can come up with, you basically know your budget.

Building Plans

Floor plans or blueprints can be obtained from either an existing set, one that the present owner may have or a set on file with local building department or historic societies, or they can be drawn for you by an architect or designer (the latter is the most likely case). You can probably sketch out the room sizes on a piece of graph paper, if it is just for your feasibility study. I recommend this method, as it doesn't cost you anything. Simply draw the inside dimensions of each room on a piece of graph paper allowing one foot for each square on the paper. You don't have to draw or sketch the whole R.O.S. as it would appear on a blueprint with each room contiguous. Just be sure to label each room as to its position and use.

Generally, you won't want or need to go through the expense of actual blueprints until you have committed yourself to the purchase of a particular R.O.S. You will need blueprints before you can proceed with actual cost estimating. They should show the structure as is, the way a sketch or floor plan would, and also how it will be when completed, showing any and all changes, additions, and deletions to the structure you plan. The blueprints will or should include a foundation plan, if applicable, floor plans for all floors, and outside *elevations*. Optional features can be a typical wall section, electrical plan, and a mechanical plan for heating and air conditioning.

If you have never looked at a set of blueprints in your life, don't worry. There is no mystery to them. The main function of residential blueprints is to show the floor plan and the design (elevation). They are virtually self-explanatory, but you can go over them with whoever drew the plans to be sure you do understand them. There is no cost for this. Be sure you understand even the items you feel foolish asking about. Your subs will understand the plans and most will use them for price quotes. A sample set of blueprints is included in appendix K. But remember your designer or architect will go over your blueprints and floor plan as many times as necessary for you to understand them.

Determining Space Requirements

Space requirements are different for different people, and what is adequate for one will be totally inadequate for another. This somewhat obvious statement has a hidden consequence, however, when you employ others to design and draw floor plans for your R.O.S. What is adequate space to them may be inadequate to you. This means living in discomfort although you may save on your renovation costs (smaller usually means less). The opposite is equally bad. You may end up with larger spaces than you need, and spend more than you need on renovation and maintenance as a result. Once again, your budget will have to temper your final decision. But how do you determine what is and what isn't adequate space for you? I find the most foolproof method is to take the floor plan and room-by-room find another room of equivalent size that already exists. There is no substitute for reality, and I strongly recommend this method. If the kitchen is 12 × 12, find a 12 × 12 kitchen in a friend's house or apartment; or at least find a room that is 12 × 12 and allow for cabinets and appliances. If you have massive furniture, allow for that in other rooms. I also recommend that you "block in" large furniture that will occupy rooms. To do this cut pieces of graph paper or plain paper to scale and place the pieces on your floor plan. To scale means that if ¼ inch equals 1 foot on your floor plan, then cut the piece of paper to show ¼ inch of paper for each 1 foot of furniture. It doesn't have to be exact, so don't fret over this little chore.

Suggestions from your designer are valuable but question his feelings of what is and what isn't adequate. The rooms that are usually undersized in a structure or floor plan are closets, baths, hallways, and kitchens.

Master Plan

I have referred before to what I call a master plan. Actually, once you prepare your budget, your cost list (chapter 5), your estimate sheet, and your priority list, you pretty well have a master plan. You know what you are and what you are not going to do. But some of the things you have eliminated need to be planned for in the future, and all the items, even though they are on the cost list, need to be written down in a general synopsis so that you can proceed in an orderly fashion. I find the easiest way to do this is to have a drawing, a sketch, or a set of blueprints for the R.O.S. in front of me. Then I can go through each room and make a list of what I am going to do to that room. I do the same thing for the exterior using the elevation. The master plan is refined as you proceed with the other planning stages, since you obviously can't decide what not to do until you find out what it costs and what the entire project costs.

Planning for Additions

If you find that you need more space than the R.O.S. allows, but everything else is what you're looking for, you can consider an addition to the structure. This is usually not difficult and may not be too expensive.

To get more space in one room in the R.O.S., you may want to combine rooms by removing a common wall or cutting a doorway, and then add the room you eliminated back on by building an addition. In chapter 5, I discuss estimating the cost of an addition and in chapter 11 the actual building of one. Additions can, and often do, use existing space within the R.O.S. that was previously not going to be used as living space; for example attics, basements, storage space, and garages or carports. Utilizing these spaces is usually the least expensive way to add on as you can take advantage of such existing structural components as roofs, walls and foundations.

Even adding a deck or porch increases, or gives the effect of increasing, living space, as well as adding value to the R.O.S. Discuss this possibility with your architect or designer. Make sure local codes (especially concerning setbacks and septic systems) allow your planned addition.

Fireplaces can be added to virtually any room in the structure, especially with today's prefab units. These prefab units are very safe, economical to install, and work well. They are also aesthetically pleasing and in many cases you can't tell the difference between a solid masonry fireplace and a prefab. They are also available in heat recirculating models for saving energy.

How Long It Will Take

The major factor in determining how long any restoration will take is the scope of the job. Obviously, if you are going to *gut* a brownstone or convert a barn, it is going to take longer than a kitchen-and-bath remodeling job. To determine how long your project will take, add up the time designated for each item in chapter 11. Use only the items that will apply to your project. (Your architect/designer can help you with this. You may or may not be charged for the time involved — *ask first.* It certainly shouldn't take more than one hour of his time or yours.) Use maximum times to be on the safe side, for it will often take that long and sometimes longer. But don't worry about that now; we will cover that in chapter 11 and the extra time is usually not important. Other factors that will affect the time it takes to complete the project are weather, availability of labor and materials, municipal inspections, and your own availability to schedule the steps, complete one, and move on to the next. As you will see, it isn't a race to the finish and you need not lose any sleep over lost time. I'll show you why in chapter 7. As you will see, minor renovation should take about a month or so and major restoration, five to eight months.

If you are already living in your R.O.S. and have no alternative but to stay during the restoration, it may be more difficult. I am not talking about minor projects now but major ones. Even a kitchen job can be quite inconvenient if you are living there. A complete kitchen can be completed in a week and if you take proper precautions you won't find it too bad. For instance, seal the kitchen off from the rest of the house — even with plywood, if necessary — to keep dust and workmen from filtering through the rest of the house. Make arrangements to take your meals elsewhere. One room at a time restoration isn't too bad but the dust must be expected, even if that particular room is sealed off. And, of course, your personal possessions must be considered, from the standpoints of security and damage. With additions, make sure that the addition is as complete as possible before cutting the doorway into the existing structure. You can have it almost ready for carpet.

△ **Before:** *Even transforming a porch into a sunspace will produce dust, debris, and potential hazards for residents.*

△ **After:** *But after clean-up and landscaping, this sunspace successfully bridges home and garden.*

△ **Before:** *Renovating a kitchen can be especially disrupting to a family lifestyle.*

△ **After:** *Here, prefinished cabinets are installed to minimize construction time, and then heavy molding is painted to match.*

△ **Before:** *Converting this once open-air porch into a sunroom is a fairly low impact project.*

◁ **After:** *And once transformed, it can be enjoyed year-round.*

In a major restoration don't move in until it is completely finished. In many parts of the country you are wisely prevented from doing so by local building codes and inspectors. Aside from the dust, dirt and lack of privacy, an incomplete project can be very unsafe. Electrical fires and shocks are probably the greatest hazard. Also, open stairways and missing railings can present potential hazards. You will also finish all the last odds and ends more quickly if there aren't the encumbrances of people and their possessions. But as always, let your own circumstances and common sense prevail.

Chapter 5

Cost Estimating

Cost estimating is not an exact science. No two people will ever come up with the same cost for a given project. I don't say this to frighten you, but rather to ease your mind. Any two people can come reasonably close. Reasonably close is five percent or less of actual cost and if you keep that in mind when you are planning your budget, you won't be unpleasantly surprised at the end of the project. You may be *pleasantly* surprised. Below is a list of items for estimating costs you may encounter in any restoration project. This list includes items for projects ranging from painting a room to a total restoration or rehabilitation, including additions. Some will not pertain to your particular project. If an exact cost is available prior to construction, I indicate where it is usually obtained in the explanations following the list.

In some areas of the country, costs by some trades may vary significantly. You may want to call several different subcontractors to verify that costs you are quoted are in line. Don't be discouraged if a few are higher. That situation exists everywhere. Some charge, and get, exorbitant prices. Just keep trying. See chapter 8 on how to find subs.

LIST OF ALL POSSIBLE COSTS FOR TOTAL RENOVATION

	ESTIMATED COST	ACTUAL COST
Inspection fees, testing (lead, asbestos, radon, and water)		
Design assistance		
Loan closing costs		
Construction or short-term loan interest		
Fire insurance		
Temporary utilities		
Cleaning the R.O.S.		
Plans and specifications		
Permits and fees		
Clearing, grading, and excavation for additions		
Footings		
Foundations		
Waterproofing		
Tearing out		
▷ exterior walls		
▷ interior walls		
▷ roof-shingles		
▷ roof rafters		
▷ plaster, drywall		
▷ furnace, radiators		
▷ plumbing		
▷ electrical wiring, etc.		
▷ total gutting		
▷ disposal dumpster fee		
▷ asbestos removal		
▷ lead paint removal		
New framing — lumber, materials, and labor		
▷ walls		
▷ floors		
▷ ceilings		

	ESTIMATED COST	ACTUAL COST
▷ roof		
▷ additions		
Steel		
Windows		
Exterior doors		
Roofing — labor and materials		
Concrete flatwork		
Exterior trim — labor and materials		
Exterior siding or veneers		
Plumbing — upfitting		
▷ toilets		
▷ bath sinks		
▷ kitchen sinks		
▷ tubs, refinished		
▷ tubs, replaced		
▷ water heater		
▷ water pipes		
Plumbing — new		
▷ kitchen(s)		
▷ bath(s)		
Heating and A/C		
Electrical — upfitting		
Electrical — new wiring and service		
Insulation		
Drywall or plaster		
Water and sewer main repair		
Well and septic system		
Cabinets, bath vanities		
Interior trim materials		
Interior doors		
Interior trim labor		
Painting, interior		
Painting, exterior		
Appliances		
Light fixtures		
Floor coverings		

	ESTIMATED COST	ACTUAL COST
Drives, walks, patios		
Cleaning		
Gutters, screens, miscellaneous		
Wallpaper		
Hardware and bath accessories		
Landscaping		
Miscellaneous allowance		

Item-by-Item Estimating

Inspection fees. These costs will vary with locality and time involved. A quick phone call will get you an exact cost.

Design assistance. Whether you use an architect or a designer, plan for approximately four to six hours at whatever the current hourly rate is in your locale. This charge does not include the cost of plans and specifications and assumes there is no supervision service included in the charge for plans and specs. It is for additional technical and design advice before and during the project that may or may not be needed.

Loan closing costs. These are charges collected by your lender and include their service charges for making the loan, *title insurance*, prepaid taxes, *recording fees*, and possibly other charges. You can get an exact quote of all charges prior to closing. I advise you to do so. Since the charges are generally based on a percentage of the amount you borrow, that amount reflects the size of your R.O.S.

Construction loan interest. As I will discuss in chapter 7, loan interest is a cost of construction and should be treated as such, whether paid on a true construction loan that will be part of the permanent note or on a short-term personal note that will be *rolled over* into a new loan when the project is done. The interest, for all practical purposes, should not exceed ten percent of the project cost, and more likely will be less. Project cost is usually related to the size of your R.O.S. and the time to complete renovation. An estimate can be obtained in advance from your lender. If you use your own money, you can record here the amount of interest you would have earned if you had invested this money otherwise.

Fire insurance. The amount carried is equal to the value of the completed R.O.S. The rate is also a function of size. An exact figure can be obtained from your insurance agent.

Temporary utilities. If there is no electricity or water available at the R.O.S., you will have to make arrangements for those utilities as discussed in chapter 10. Portable sanitation units also fall into this category. The sooner you can ready the plumbing in your R.O.S., the sooner you can eliminate this cost.

Cleaning the R.O.S. Before you start renovating, cleaning is an important job as you'll see in chapter 10. This is sometimes a job that you can do, but if not you can get an exact quote from a professional service listed in the Yellow Pages.

Plans and specifications. An exact quote can and should be obtained very early in the game from your architect-designer.

Permits and fees. These are paid to local government agencies. You can get a quote over the phone. See chapter 10 to find the proper information.

Clearing, grading and excavation. As seen in chapter 11, this function generally applies to additions. An exact quote can be asked for beforehand.

Footings. For an addition, or when adding footings for a structure that doesn't have them, such as a barn. This includes digging the trenches, forming *rebars* (if necessary), purchasing the concrete and pouring it. It is based on an average 8 inch × 16 inch × 8 inch footing and does not include removal of old footings or jacking a wall up in order to place a footing under it. Those two items would have to be priced by getting quotes from the appropriate sub (see chapter 8).

Foundations. This cost doesn't include demolition of existing walls to be rebuilt. It does include all materials — block, brick, stone, sand, mortar, Durawall — and all labor. For double walls, double the estimate. If you can't determine proper square footage, have your designer, or supplier do so (free!). Some brick, block, or stone masons will give a contract price, others will charge by the brick or block. Stone masons generally charge by the square foot.

Waterproofing. Use professionals!

Tearing out. Prices are usually quoted by the square foot from your carpentry contractor. Disposal fees are significant in most areas; dumpsters are available in 10 and 30 cubic yard capacities. Special disposal regulations may be applicable for hazardous wastes such as asbestos.

Framing — lumber, materials, and labor. Prices usually quoted by the square foot from your carpentry contractor. You can also obtain a materials cost from your lumberyard and a labor only quote from your carpenter. Usually this item includes all items to enclose the house and make it reasonably weathertight ("drying in") except for doors, windows, roof shingles, and siding.

Steel. Steel beams and lintels may be required where there is serious sag or a large span. It is priced by the linear foot.

Windows. Windows will vary in price depending on options such as glass type (insulated glass is standard and required by most codes) and options such as grids, cladding, sash locks, and screens.

Exterior Doors. Wood doors are the most expensive, followed by insulated fiberglass doors and then insulated steel doors. Prices quoted from your supplier will include pre-hung units, weatherstripping and jambs (with hinges), but not handle or lock hardware. Your carpenter can give you an installation price per unit. This cost should allow for installation of pre-hung units and weatherstripping, but not hardware other than hinges.

Roofing — labor and materials. Roofing is quoted by the "square," which equals 100 square feet of roof area. If you plan on removal of and hauling off old layers, double the labor charge. Old roofing requires a lot of dumpster space; if the existing roof is asphalt and lays flat it can be roofed over twice above the original layer without removal. The steeper the roof, the higher the labor charge will be. Get an exact quote from a roofing subcontractor, and make sure they include all necessary flashing, especially around chimneys.

Concrete flatwork. These are slabs used on garage and basement floors where concrete is to be finished smooth. It also involves use of insulating and reinforcing materials as required by municipal codes.

Exterior trim. These are the materials, including: soffit, fascia board, *soffit vents*, window and door moldings, and posts. Different houses will require different items, and a list of them can be made by a supplier or designer and an exact cost predetermined.

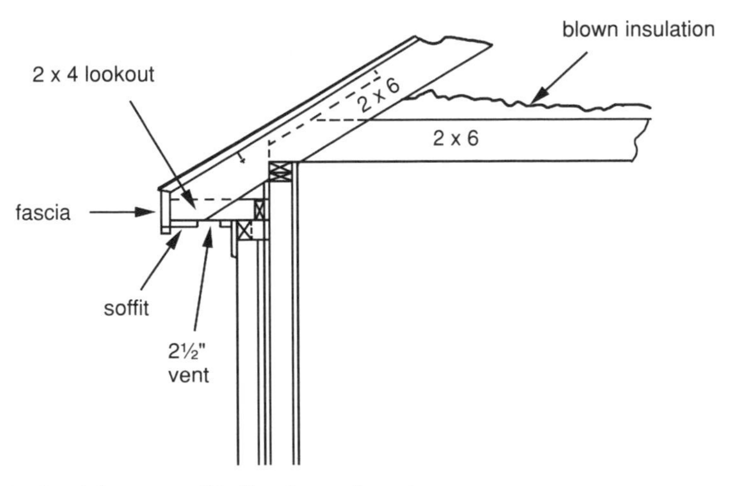

Exterior trim — soffit, fascia, and vent

Exterior siding or veneers. These prices are quoted by the square foot of surface area to be covered and can be obtained from your supplier or subcontractor. Materials can range from vinyl to brick. Old siding is generally removed and an insulating sheathing board installed under the new siding, greatly improving energy efficiency. For asbestos siding, it's best to install the insulating sheathing directly over it and your new siding over that. This helps to contain the asbestos if it cracks and is much less expensive than the special disposal costs associated with asbestos removal.

Plumbing — upfitting. Plumbers will quote a per fixture installation cost, plus the cost of the fixtures. Plumbers can usually give you a better price on fixtures than if you shop for them yourself. Note that new codes may require low water usage fixtures. This quote should include sinks, water closets (toilets), tubs, showers, laundry hookup (if you don't have one), and water heater, but not appliances such as dishwasher, washing machine, or garbage disposal.

Plumbing — new. This will include all plumbing, fixtures, and installation (except those kitchen items that are considered appliances such as dishwasher, disposal) where the cost of plumbing them only is included. Relatively little plaster or drywall needs to be cut out for full plumbing restoration. Patching is therefore relatively easy.

Heating and A/C. This estimate would include all new ductwork, furnace(s), compressor(s), and thermostats. I prefer a forced-air system with A/C (which is comparable in price to a heat-pump), with gas or oil as fuel. Hot water baseboard systems cost more to install but cost less to operate. Electric heat costs the least to install but the most to operate. Gas, oil, or electric without A/C is less expensive to install. Heatpumps have A/C as an integral part of their operation so you get A/C automatically. Oil or gas boilers for water systems should be quoted individually, unless you choose a hot water baseboard heating system. New boilers can heat your hot water for your faucets and also for your baseboard space heating system in one unit. If you want central A/C, a forced air heating system becomes more attractive economically. Exact quotes can be asked for early in the game. Vents for bath exhaust, dryer exhaust, and range hood can be handled by your electrician who can also install the fans. (Be sure all vents are vented to the outside, not to attics, basement, garages, or crawl spaces.) Ventilation systems to provide fresh air (separate from your heating system) may be needed if the house is being made very tight with new insulation, or if you need one for getting rid of a radon problem.

Electrical — upfitting. Electricians will usually quote a price based on each device to be wired and the linear footage of wiring to be run through existing walls. Your service may have to be increased because of the heavier loads required by new appliances, and your service panel may have to be upgraded from fuses to circuit breakers.

Electrical — new. Estimates based on all new switches; receptacles; circuit panels and breakers; wiring of all built-in appliances including water heater, furnaces, and A/C; and wiring of all bath fans (cost of fans is under light fixtures). An exact quote based on your blueprints can be obtained from an electrician. Total new wiring can be done without removing all the plaster or drywall. A few cuts or holes in each wall may have to be made, but patching is relatively easy.

Insulation. Amounts of insulation required in walls, floors, and ceilings will often be dictated by current building codes. For example, R-13 may be required in the floor, R-19 in the walls, and R-30 in the ceiling. Price for installation will vary depending on accessibility (good if your R.O.S. has been gutted first) or if holes have to be drilled, insulation blown in and holes plugged in exterior walls. For your needs, which will vary with climate, consult an insulation contractor and your local utilities.

Drywall or plaster. Drywall (gypsum wallboard) or plaster is quoted by the square foot of surface to be covered and includes hanging, taping, and sanding. Estimates for a complete job can be obtained from a drywall sub. Plasterers may be harder to find these days, but new types of gypsum board have been developed to accept "skim coat" plaster. Suppliers of this material can usually recommend a plasterer.

Water and sewer main repair. If these systems are fifty years old or more replace them as soon as feasible.

Well and septic system. Where water and sewer are not available, and where conditions permit obtain a firm quote for both.

Note: Some well drillers can give a contract price, especially if they are familiar with the area. Shop carefully. Try and avoid a per foot drilling charge. Be sure the well and its water meet local codes and health department requirements before paying for it.

Septic systems will vary by local requirements. Bedrooms determine the number of people living in the structure and thus how large the system, and price, should be. Again, be sure the system meets all codes and has been inspected before paying. If your locale doesn't have a building inspection department or health department inspector for either well or septic systems, consult your inspection engineer. He's qualified to inspect and any small additional charge is well worth it.

Cabinets, bath vanities. An exact quote per linear foot can be given by your supplier; price will vary drastically. An average kitchen will have 12 to 15 feet of cabinets. Cabinet quotes should include labor to install and should be for a finished (painted or stained) product, including all hardware.

Interior trim materials. Estimated per linear foot of each room.

Interior doors. Costs include prehung units with casing. Solid wood doors are the most expensive; hollow core hardboard the least.

Interior trim labor. Labor includes applying standard trim and hanging all doors.

Painting, interior. Square foot cost includes all materials (2 coats of paint). Exact quote from the painter may be obtained, using the plans.

Painting, exterior. Square foot costs include all materials and caulking of all seams and joints, exterior painting. Extra for sanding and scraping. Exact quote from the painter may be obtained, using the plans.

Appliances. Costs will vary widely by brand and quality.

Light fixtures. Quotes from light fixture suppliers can be taken from your plans. The sky's the limit. Most old existing fixtures in an O.S. can be considered unsafe and should be repaired or replaced. Repairs should be made by an Underwriter Laboratories (U.L.) approved technician listed in the Yellow Pages. Local codes may require this.

Floor coverings. Hardwood (oak) flooring runs $2.50 to $3.00 per square foot, installed, sanded, stained, and sealed with polyurethane; $1.00 per square foot for refinishing only. Prefinished parquet costs $3.00 to $4.00 a square foot; carpet, $9.00 to $50.00 per square yard; sheet vinyls, $8.50 to $18.00 per square yard; ceramic tile, $5.00 per square foot and up; slate or brick, $5.00 per square foot.

Drives, walks, patios. Use concrete 4 inches thick with a broom finish.

Cleaning. Same as item 7, but this time more thorough, including windows. Average cost can run as high as $.15 to $.20 per square foot.

Gutters, screens, miscellaneous. Gutters and downspouts run approximately $1.50 per linear foot for aluminum, screens are approximately $10.00 each, and miscellaneous makes up the difference. Miscellaneous is anything not specifically planned for and these items, such as garage doors, should be accounted for.

Wallpaper. This is a non-necessity, but most builders give their customer a dollar allowance of approximately $350 to $500 installed. Why not use that?

Hardware and bath accessories. Sky's the limit.

Landscaping. Stonework, lawns, trees, shrubs.

Miscellaneous allowance. Figure on $4,000 to $6,000 or five percent of the total cost, whichever is greater.

Ballpark Estimating

Obviously completing all items of estimating will take some time and effort on your part and it's not feasible to do it for every R.O.S. you look at, only when you get close to buying a particular one. So what do you do, or what do you use to enable you to estimate quickly when you are in your search for your R.O.S.? You use "ballpark figures" to aid you in your decision-making process. It is not very accurate, but accurate enough to decide whether you should take the time to do a cost estimate on it. This procedure concerns itself with the basics and the most costly items in restoration. It's mine and it works for me. All figures are based on an average-sized house.

- ▷ Total gut and restoration — all you have is four walls, floor and ceiling joints, roof that needs repair, and a fairly sound structure when you're done — $1.50 to $2.00 per square foot to gut and $30 to $40 per square foot to gut and restore completely
- ▷ New kitchen — (average size) complete including tearing out — $6,000–$8,000
- ▷ New bath — each, complete and including gutting — $3,500 to $4,500
- ▷ New roof — $2,000 to $5,000
- ▷ Repair plaster — $1,000 if in fairly good condition; cover plaster with drywall — $3,000–$4,000
- ▷ Paint — complete, $2,000–$4,000
- ▷ New electrical — complete, $3,000–$4,000
- ▷ New heat-A/C — $4,000–$5,000
- ▷ Insulation — $4,000–$6,000
- ▷ Extensive carpentry — $2,000 to $3,000

Estimated vs. *Actual Costs*

As you look at the estimate sheet on page 55–57, you will notice two columns; estimated cost and actual cost. The estimated-cost column is derived and filled in as discussed at the beginning of this chapter. The actual-cost column is filled in as you pay for each completed item, and not until then. Even though you may have a firm bid or quote from a sub or supplier, you are never certain of actual cost until that particular item is complete. Change and add-on items can change firm bids and quotes. For example, your plumber may quote a total price of $2,500, including all fixtures. After work starts, you may change your mind on the style of the commode and could conceivably add $200 to this quote. As you complete each item, pay for it, and enter it into the actual-cost column, you will be able to see how your project is coming out cost wise. Early attention to any cost overrun, actual more than estimated, is advised. Cost underruns are to be taken with a grain of salt as they most likely will be eaten up by another item's overrun.

Major and Minor Costs

To avoid letting your project end up with serious overruns (more than five to ten percent), an understanding of major versus minor cost is imperative. The major costs are those listed in the "ballpark estimate." All the others should be minor in cost, or cosmetic in nature and therefore discretionary, or items that can be done later.

Chapter 6

Buying the Restorable Structure

The **planning stage** of your renovation project consists of two phases. The first phase is *before* you purchase your R.O.S. During this time you are determining whether the structure is restorable, whether the restoration is feasible for you, and what the cost of the undertaking will be. The second phase of the planning is *after* you have purchased the R.O.S. During that phase you will actually be selecting, hiring, and scheduling subcontractors; setting up accounts to purchase supplies; and making arrangements for miscellaneous matters concerning the upcoming renovation work.

But in between the two planning stages comes the important matter of purchasing and receiving financing for the R.O.S. I will discuss financing in the next chapter, but now, having roughly determined the cost of renovation, how do you determine the present value of the R.O.S.?

Pricing the R.O.S.

The most common method of determining how much any structure should sell for is by the use of comparables. Notice I said "sell for," not the asking price. Comparables are structures that have recently been sold (within six months) and are similar in nature, and in the immediate neighborhood. For example, if you are looking at a row house of approximately 1,200 square feet and nearby three or four of like size have sold within the last six months for approximately $40,000, you should expect to pay about the same. This is, of course, keeping all things equal, including primarily the condition of the structure, and a not too active demand for the structure(s).

There are several ways of finding comparables; the easiest is to use a real estate broker. Your broker has immediate access to all previous sales information through a multiple listing service, and the register of deeds at the local courthouse. If you do not use a real estate broker (and I do *not* recommend this), you still have access to the register of deeds. This is the best place to discover how much the comparable sold for. You can tell which ones have been sold recently by studying the area and seeing which houses are recently inhabited or if new owners have moved in. Then get the address and look it up at the courthouse. However, not all locales record accurate sale-price information, so be sure to ask at the courthouse how such information is compiled and recorded. If you are using a real estate broker, he or she should do all of this for you. The general condition of an R.O.S. and its neighborhood are the determining factors for setting sales price, and as I mentioned earlier, as the neighborhood improves, demand goes up, along with asking prices. On a first time purchase, I can't conceive of not using a real estate broker for such an important step.

What if no comparables exist? For example, an old barn in the country with nothing but farmland around it? In this case you or your broker use acreage as a comparable and analytically determine how much dollar value the structure has by itself. If you are only getting a foundation, then that's all it is worth. See chapter 5 for the cost of a foundation. If all you are getting is a foundation, roof and walls, then that's all that is worth. Determine their cumulative worth from the estimate sheet in chapter 5. If those items you are adding up are in need of repair, be sure to deduct that amount from the worth you assign to the structure. In other words, if you are considering an R.O.S. in which half of the windows are broken, look at the estimate sheet and calculate the cost for all of the windows. This is what the windows would be worth if they were all intact. Deduct the cost of replacing the broken windows from the worth-if-intact to get the present worth of the windows.

When you have finished, you will have a ballpark estimate of what that particular R.O.S. is worth. What it will sell for depends on how badly the owner wants to sell. There is an old saying that is particularly true in real estate — anything is for sale for the right price. What is "right" for the seller is seldom "right" for the buyer. The more the seller needs or wants to sell, the closer to your "right" price you will come. This all evolves in what is known as the "offering process."

There are, however, cases where no offering process takes place and the asking price is the selling price. This is typical in government sale of structures and foreclosure sales by lenders (although sometimes in the latter case and on rare occasion in the former, there is some room for negotiation).

There is also the case where the owner doesn't need to sell and so will only sell at his or her asking price, which you can safely assume is too high for what you are getting. But, in the vast majority of real estate transactions, a little game takes place that produces an end product — the selling price. The game is called the offering process and is best played by a competent referee — your real estate broker, or even the seller's broker. It can be a tricky game and it is one you certainly don't want to lose. If it is played properly and fairly, it is probably the only game where no one loses.

The Offering Process

It didn't take me long to realize that I priced my new houses too low. When someone pays the asking price for a piece of real estate, it is either priced too low or the buyer is acting in an unbalanced manner.

I learned, as most sellers learn, to ask much more than I expect to get, in hope that I might get it (there's more than one unbalanced soul out there)! If that sounds devious or unfair, it is a fact of life and not one that I created. It exists in almost every facet of retailing. (Know anyone who paid sticker price for a car? There are those who do.)

The game of offering has no firm rules but proceeds when you, or your real estate broker, offer the seller a price so much lower than the asking price you know he's going to be insulted. If your broker says "We don't want to insult the man," simply say, "Sure we do!" An insulting offer to a seller often knocks the wind out of his sails and quite often will start him thinking that perhaps he is asking "a little too much." Then the seller will counter-offer with a slightly lower price to test the water, and you counter *that* counter with a slightly higher price, and so on until one or the other of you gets so mad that someone will say "forget it" (at which point, you start over again) or you reach a "deal" — a selling price. In the middle of all this madness, it really does help to have that referee, or a buffer so to speak, giving advice and guidance — a real estate broker. Besides, a broker is familiar with concessions that are available other than a lower selling price, mainly seller financing which is better than money. We discuss seller financing in more depth in chapter 7. Any final offer to buy should contain any *contingencies* you and your broker feel necessary for your protection. They should at least consist of:

- ▷ An acceptable inspection of the structure
- ▷ Sufficient time for you to do a complete cost estimate and arrange financing
- ▷ Time to consult with an architect or designer to determine feasibility of restoration
- ▷ Anything else you or your broker deem necessary

Working with a Real Estate Broker

You should see by now that your need for a real estate broker is imperative for this important early step in realizing your dream. His or her fee is earned and justified and, as we have seen, can often save you more money than the amount of his or her fee. You must, however, when dealing with a broker, remain in control of the relationship and not be intimidated. Find one you like and feel you can trust — by intuition and reference checking — and spell out your needs clearly. If you fail to get satisfactory results in what you consider to be sufficient time, find another. Most brokers are accustomed to working for the seller in that they are usually *pursuing* listings. Be sure they understand that in this situation, finding your R.O.S., they are working for you and are to represent *your* best interest.

Another important point concerning real estate brokers you should be aware of is that brokers are often in the business of finding and restoring O.S.'s for themselves or local builders. I would strongly suggest you avoid using such a broker as the better R.O.S.'s are likely *not* to be shown to you.

Your familiarity with the value of an R.O.S. is as important — perhaps more so — when working with a broker. The ability to quickly size up value will come, I assure you. It's like looking at used cars, but with fewer potentially hidden problems. And don't feel that you are inconveniencing your broker by wanting to see as many R.O.S.'s as possible. Stay in control, demand the best service, or get a new broker.

Quite often R.O.S.'s are not advertised for sale or listed with a broker. The owner may have been thinking about selling, but hasn't yet decided to do so. If an R.O.S. that you find appealing is occupied, you can knock on the door and after introducing yourself, ask if it is for sale. If it is occupied by a tenant, they can inform you of the owner or agent from whom they are renting. Or you can look up owners at the tax office, and call or write to let them know you *may* be interested in purchasing their structure.

Perhaps it is a store with the upstairs vacant. The owner, if not planning a different use for the building, should be interested in converting that wasted space upstairs to an apartment or condo but hasn't thought about it, or has been putting it off. A phone call, letter or personal visit by you or your real estate broker could precipitate such a transformation.

Another tried and true method of finding unadvertised R.O.S.'s is for you or your real estate broker to ask people in the neighborhood if they are aware of any, and if not, if they would keep their eyes open for one for you. If you are using a broker for this process I recommend that you accompany him as it is an excellent way to meet and evaluate your potential neighbors.

Put Everything in Writing

If you remember only one thing about real estate, I urge you to remember this: all transactions concerning real estate (land, building, rental) absolutely must be in writing. An oral agreement to buy or sell is worthless and totally unenforceable if either buyer or seller waivers. Going one step further, before you sign an agreement I urge you to have any written documents, even those standard forms from real estate brokers, reviewed by a competent real estate attorney. That is, an attorney who specializes in real estate. And, yes, there are incompetent ones. See chapter 8 on how to find a good one. You will need an attorney to close your transaction anyway. Your own attorney is preferable and chances are if your attorney *is* going to close the transaction he won't charge to review it first. I wouldn't think of operating in this business for one second if I did not have a good attorney. He is just another one of the key people that I employ and rely on.

Chapter 7

Financing

Financing the purchase and restoration of your R.O.S will be easier than building a new house if you act as your own general contractor. The reason is, you are starting with something. You will have an existing structure on which a lender will grant a loan for at least a percentage of its existing or present value. This seemingly trivial point is important because in new construction you have to convince a lender that you can complete the construction of a house. With an R.O.S., you already have a structure, often livable as is. The lender can thus lend money for it. The *collateral* they need already exists, at least in part. In this chapter, we will see how to finance both the R.O.S. and the restoration. Obviously, this whole chapter assumes you do need financing, that you are not paying cash.

There is an old rule of thumb in the housing and lending industries, that your monthly mortgage payments, including taxes and insurance, should not exceed 25 to 28 percent of your gross monthly income, from all sources and, if applicable, both spouses. That rule flew out the window a few years ago in California and the change has been flying eastward ever since. It is not unusual for total monthly payments to require 40 percent and even 60 percent of a family's gross monthly income. Lenders frown on this, but with owners financing parts of the sale and other types of secondary financing (which we will cover later), they are often unaware of how high your total monthly payments are. Unless you are cautious you can actually end up spending upwards of 60 percent of your gross monthly income and *you* won't even be aware of it! I think this is a very dangerous situation and I hope you do, too. It leaves very little for the other necessities and pleasures of life. This situation is known as being "house poor" and because you have more house than you can afford, you have little enjoyment other than knowing you have the home.

There is often a tendency for those who act as their own general contractor to feel that since they are saving so much they should put that savings into a larger house. That is fine *if* you can really afford it, but I recommend you use the savings as your down payment or to lower the amount you will need to borrow for a finished product that you can easily afford. I recommend that you don't go above 30 percent of your gross monthly income and, by managing your own renovation project, you shouldn't have to. Each locale and each lender varies, so *before* you buy an R.O.S. sit down with several lenders and determine how much home you can afford. You don't have to, and at this point probably shouldn't, tell them you are going to manage the project yourself, or for that matter, that you are going to restore an old structure. You only want to determine how much you can afford. You may as well find out from the expert, and while you're there you can also explore a few other matters we will discuss in this chapter. All the information and the time spent with the potential lender is free.

What You Save Can Be Your Down Payment

I've said that the money you save by managing your own renovation project can be used as your down payment. This will depend on how much you are going to save and how much your R.O.S. costs you in its existing state. All savings are based on the structure's market value when it is restored. For example, if you were to purchase a 2,000 square foot house for $20,000, and it was going to cost $30 per square foot to restore (that would be a full gut job), and it would be worth $50 per square foot when you were finished, you would have saved $20 per square foot by managing your own renovation project — an amount equal to a 20% down payment.

Purchase price of R.O.S.	$ 20,000
Restoration cost (2,000 x $30)	*+60,000*
Total cost of R.O.S.	$ 80,000
market value (2,000 x $50)	$100,000
Total cost of R.O.S.	*- 80,000*
You save	$ 20,000
Normal down payment 20% x $100,000	$ 20,000

Now that means the lender will lend you $80,000 to buy the R.O.S. and restore it. We will cover how the lender gives you these monies ($20,000 to buy plus $60,000 to restore). (The lender doesn't just give you a lump sum of $80,000 and say "Go to it. Call me when you are finished!")

What Lenders Require in Working Out a Loan

First let's discuss what lenders will require before they lend you money to manage your own renovation project. If you were to simply ask for a long-term loan, a mortgage, just to buy an old but livable house, without any additional money for restoration the lender would have no hesitation, provided:

- ▷ The R.O.S. has intrinsic value as is
- ▷ Your credit and salary justify such a purchase
- ▷ The R.O.S. is habitable as is, perhaps in need of only minor or cosmetic repairs that are not urgent

But, if they are going to lend you money to restore the R.O.S. they are going to want assurance that the project will be completed and at or near its estimated cost. Past incompleted projects or those that incurred large cost overruns make them wary. They will be wary if *you* manage your own project; after all, they are even wary of professional general contractors.

If you call a lender and ask if he will lend you money to restore an R.O.S. with you acting as your own manager, he would most likely say no — and quite justifiably so. He doesn't know whether you can do the job and doesn't want his company to have to step in and finish it for you. An automatic no is usually a matter of policy in your initial talks with a lender. They will, however, on a case by case basis allow borrowers to manage their own renovation projects if they meet the following two criteria:

1. You have planned well in advance and exude confidence This planning should at least include:
   a. Finding a suitable R.O.S. at a realistic price
   b. Restoration planning and cost estimating, schedule, and drawings (architectural plans)
   c. Lining up a few major subcontractors
   d. Lining up a few major suppliers (the easy ones)
   e. An understanding of the procedural process of restoration (That, I hope, you will learn from this book.)
2. You qualify for the dollar amount of the loan desired. That is, you can afford it

If you choose not to attempt steps a. through d. until you have first found a willing lender, you will seem less impressive to the lender. And remember, your offer to buy an R.O.S. should be contingent on being able to find suitable financing. If you can't get the loan, you are not obligated to buy. You will, however, have spent a few hundred dollars and some of your time. Remember, you do have an alternative if you can't find a lender willing to allow you to manage your own project. That is the "manager's construction contract" with a professional general contractor, as discussed in chapter 2.

By all means, be persistent in finding a lender. It's a selling job; it is yourself you are selling. You'll feel more comfortable after you break the ice, and more at ease in dealing with lenders. But if you really have the confidence that you can do it, let that translate into enthusiasm in your talks. That is what selling anything is all about, and after all you really do have something to sell — yourself. Selling is friendly persuasion. Ten years ago my request for a construction loan was turned down by two different lenders. I finally realized what I had to do was *persuade* a lender to make me that loan.

The lender's reluctance to lend is due to the unfinished portion of your project, that is, the money needed to restore. This money, loaned as short-term financing for twelve months or less, is the construction loan. It is money committed by the lender for something rather intangible; that is, a set of plans and your promise to finish the O.S. in accordance with those plans.

The money is committed at the onset or closing of the purchase of the R.O.S., but it is disbursed to you to pay subcontractors and suppliers only as work progresses.

The normal procedure, using our row house example outlined above, would begin with a commitment by the lender to lend you $80,000. The deal would be closed as follows: The lender would provide you an advance of $20,000 to purchase the structure, and hold the remaining $60,000 to be disbursed as construction progresses. A typical inspection report and disbursement schedule is in appendix G. It should be noted, and your lender will also explain this, that you pay interest only on the monies you actually receive. That means the first month, you would pay interest on the $20,000 plus whatever amount was advanced during that month for construction costs. This is an important point. Many people think that they will be paying interest on the entire amount ($80,000 in this case) from day one. Not so! As your renovation progresses and is approved by the lender on inspection, parts of your loan will be disbursed to you. When the project is 20 percent complete, 20 percent of the balance ($12,000) of the construction loan will be disbursed to you, and the following month you will pay interest on that additional $12,000 as well as on the unpaid balance of the $20,000. It is always the lender who makes the decision to disburse monies. Some lenders will not pay the first *draw*, or part, of construction monies. They may want you to pay the first portion of the renovation costs and reimburse you when the project is completed. You will still save but you'll have to wait until the project is finished, when you receive your permanent loan. I would recommend such a lending arrangement only as a last resort, especially if you will have difficulty paying that first $20,000 purchase cost. Since you will only need this $20,000 for a short time, you may be able to finance it with a loan from a relative, as an unsecured note from a commercial lender, as a second mortgage on another piece of real estate or through another short-term borrowing plan.

When the project is completed, the construction loan is converted to a permanent loan, the terms of which you and the lender have already agreed on. If you had to put money up front as described above, this is when you will get it back. For example, when finished you will have spent $20,000 to purchase the house and $60,000 on materials and labor, but you will only have received $60,000 of the loan for $80,000. When the renovation is 100 percent completed you will get your last draw of $20,000, before the construction loan is converted to the permanent loan.

Since all materials and labor are paid for, the $20,000 is yours. These figures are hypothetical and it should be pointed out that it rarely, if ever, works out exactly.

Loan Alternatives

Permanent loans are changing and dying in today's volatile money market. Whatever I write here today could be obsolete information by the time you read it. Do not worry. The staff at your local savings and loan institution, your real estate broker or a mortgage broker can bring you up-to-date when the time arrives.

A myriad of mortgages exist today. A conventional loan was the best, and is the fastest disappearing. It offers a fixed rate of interest for the longest available amount of time, twenty to thirty years. Unfortunately, the fixed-rate is prohibitively high for most, unless you are (as you will be) saving a considerable amount of money to help compensate for that high interest. There are also variable-rate mortgages; graduated-payment mortgages; Federal Housing Authority (FHA), Veteran's Authority (VA), and Farmers Home Administration mortgages; home equity loans and other forms of relative financing, for example, *wraparounds*, *lease options*, and growth equity mortgages *(GEMs)*.

The fastest growing type of financing today is "seller financing." Often because they have considerable equity in a property and a need to sell, owners will finance part or even all of an R.O.S. at a lower rate than conventional lenders. Generally their terms aren't as long, but surprisingly the payments are about the same for one-half to one-third of the time because of lower interest rates. There are drawbacks to some forms of seller financing. You need to discuss all ramifications of each form of financing with your real estate broker, your attorney, and your lender. The seller will not, of course, finance the restoration so you will still need a construction loan. Another good example of your need for a real estate broker emerges here. The hardest part of your project will be behind you when you have arranged *all* your financing. It is the most taxing part of the project, mentally, if not physically.

Chapter 8

Subcontractors

Now begins phase two of your planning. The groundwork you did in determining feasibility and in cost estimating will be invaluable to you now, as you arrange with subcontractors for the actual work to be done, arrange for them to purchase supplies or purchase them yourself, and plan for such miscellaneous matters as complying with building codes, providing temporary utilities, and insuring the R.O.S.

All the skilled labor jobs that comprise the renovation project are done by subcontractors. A firm or an individual contracts with you to perform a service for a predetermined price. Under no circumstances should you deviate from this procedure. That means *absolutely* no one is to do any work for you by the hour or at an unspecified price. It doesn't matter how small the job. The only exceptions are those who earn consulting fees — architects, attorneys, or inspection engineers — and then only at a previously agreed hourly rate and time estimate. Deviation from this rule will almost certainly lead to cost overruns and conceivably quite large ones.

Also, by subcontracting all labor services, *you* will have no employees and will avoid all the governmental red tape that having salaried hourly employees requires.

Kinds of Subcontractors

The following is a list of most if not all, of the subcontractors you will be dealing with and in the general order you will need them. A brief description of their duties is included where I feel necessary. Some have already been discussed. You may not need all of them.

1. Architect or house designer
2. Lending officer at savings and loans or other commercial lending institutions
3. Inspection engineer
4. Real estate attorney
5. Carpentry contractor
6. Grading and excavation contractor
7. Footing contractor
8. Masonry contractor (for foundation and chimney repairs or additions; also for masonry veneer repair and additions)
9. Concrete contractor (for basement and garage concrete-slab floors; also for drives, walks, and concrete patios)
10. Waterproofing contractors (or waterproofing foundations)
11. Electrical contractor
12. Plumbing contractor
13. Heating, air-conditioning, ventilation contractor (HVAC)
14. Roofing contractor
15. Insulation contractor
16. Drywall and/or plaster contractor
17. Painting contractor
18. Flooring, carpet, formica contractor(s)
19. Tile contractor
20. Cleaning contractor (may also be needed very early in project)
21. Landscape contractor

Finding Your Subs

I have found that subs who are primarily involved with new construction are more competitively priced than those who specialize in remodeling. This is because cost-conscious general contractors keep bidding competitive among these new construction subs.

Your first sub can lead to others, and the recommendation of one by another is a very good starting point. To find your architect/designer, you can ask lenders or real estate brokers or refer to the Yellow Pages. Carpenters can be found by asking at lumber suppliers or by stopping at a job where new construction or restoration is in progress. Grading and excavation contractors are best found either on a job-site, listed in the Yellow Pages, through suppliers, or by asking the other subs as you line them up. Most masonry, concrete, electrical, and plumbing subs know at least one other sub in each of various trades. The exceptions may be waterproofing, flooring, carpet and Formica, and cleaning contractors, but these can best be found listed in the Yellow Pages. Many good, low-priced subs are not, however, listed in the Yellow Pages. Signs at job sites can be a good source of information. By stopping to see or calling someone who has a sign posted on a tree, pole or truck, you can ask for name of non-competing tradesmen. For the most part, subs beget subs!

You can also ask owners of completed projects for subs. They usually know the subs who did the work on their house, even if they hired a general contractor. If they are happy with the work, you get a reference and a name in one swoop. "Did the electrician come back for minor adjustment?" is a typical question to ask an owner. You can also, provided you have enough moxie, call a general contractor and ask for sub recommendations. "I desperately need a good carpenter. Can you recommend one?" It works; try it if you need to.

The most important aspect of finding good subs is to get and check references. Even check references on your architect and attorney. If they take offense move on. I recommend at least three references on each, and in the case of conflicting references on each, get five. It only takes a minute or two to telephone a reference. "Were you pleased with Joe Smith's work and was he reliable?" is all you need ask. If I had a choice between quality and reliability, I'd choose reliability. You hope you won't have to choose between the two, but if a man doesn't show up when he says he will, or doesn't come back to finish a job when he is supposed to, what difference does it make if he's the best darn carpenter east of Hawaii? The same applies for the more professional subs, architects and attorneys. If they don't or won't return calls, who needs them? There are good, reliable people out there and finding them is not hard, but a little time and energy are required on your part. That is precisely what you, acting as your own general contractor, are getting paid to do.

Subcontractor Bids and Hiring

Some subs supply materials and include those material costs in their bid prices. Some do this as a matter of tradition, others do it if you ask. A carpenter does not supply lumber except on an extremely small repair job. A brick mason often supplies bricks, blocks, sand and mortar. An electrician's bid always includes all wiring, devices (switches and receptacles), breakers, panel boxes, installation of all major appliances, and wiring of furnaces, air-conditioners, and water heaters. The electrician's bid can also include such items as bath fans, exhaust vents, eaves lights, and utility lights for attics, basements, and crawl spaces. Be sure you specify in your plans, specifications (see appendix J), and written contract, what is to be supplied. You can only compare one electrician's bid with another's if they are exactly the same. Plumbers include everything right down to the toilet seat. If you haven't already selected plumbing fixtures, then use a moderate price for bid purposes with the understanding that if you upgrade to a more expensive line you will pay the difference in cost.

Many subs today are supplying all materials in their bid prices. I recommend this practice. It should cost the same as if you bought the materials, but it's one less chore for you. The more completely "turnkey" a bid is, the less work and time for you. You won't have to schedule material delivery, open charge accounts, make orders, or pay for materials when the sub supplies them. Ask each sub exactly what the bid (price) includes, and ask if materials will be supplied. Just about everyone will supply materials. For your first project I recommend getting at least three bids for each item. If you get three bids on an item and all are too high, or too low (realizing that prices in some areas of the country are much higher for some trades), or if your architect thinks the estimates are too high, or if your neighbor paid much less for a comparable job, then get more bids. If two out of three bids are close together, choose the lower of the two that are close. Some subs prey on the uninformed — especially in restoration work. You are no longer uninformed.

Some subs charge too low a price and won't be in business to service "call backs" ten months from now. Use common sense. All bids should be in writing and can be as simple as the one shown in appendix D or as intricate as you and/or your attorney deem necessary. Trust nothing to memory; put it in writing. Also, be sure to have terms of payment either spelled out in writing, or at least verbally agreed upon — preferably the former as you'll shortly see. If a sub uses a word or term that is not clear to you ask him to explain it. Don't be afraid to ask questions. In the end you'll be respected for it.

Scheduling Subs

Subs should be scheduled according to the sequences given in chapters 10 and 11. In most cases, as I have indicated in these chapters, more than one subcontractor can work at once. Subs will generally assist you in scheduling for they need the work and, in most cases, don't get paid until they are finished. Remember that their reliability is extremely important. The reliable ones will cooperate to meet your scheduling needs. I have a crew of subs that coordinates work time so that they are not in each other's way but can complete the jobs as quickly as possible. The electrician likes to be there after the plumber and the heating and air-conditioning people (he doesn't like to take the chance of someone accidentally cutting his wiring), so he learns from these subs when they will be finished. You may not be lucky enough to find such a cooperative group the first time, but then again you might.

Your single most effective control over your subs, including your architect, building inspector, and savings and loan inspector, is *how* you pay them. If you have procured reliable ones they will most likely do their best work anyway, but your purse strings are still your control.

No one is ever paid in advance for anything in this business. Advance payment lessens the urgency of some to complete or even start your project. It is permissible to pay for work-in-progress for some of the trades, but only for the percentage of the work completed less than 100 percent. This is called a "draw." Many subs require a draw if they don't have enough working capital to sustain themselves and their crews until completion of your job. This is fine. Be sure to discuss this beforehand with each sub and don't be embarrassed to do so.

Carpenters are usually paid during each stage of their work; tearing out, framing, interior trim, exterior trim. If one carpentry crew is to do all stages, you can pay for each as it is completed, less 10 percent hold-back. This hold-back is assurance that if you have to hire someone else to complete the work you will have enough money to do so. It will cost a premium to have someone else finish another's work or contract. Plumbers and electricians usually require a draw of 60 percent of the contract price when their rough-ins are finished, but the draw should not be paid until the work is inspected by a municipal inspector or an inspection engineer. Draws for HVAC contractors vary according to the amount of equipment to set in place before they *trim-out* the job. It should never exceed 60 percent unless all equipment is set, including any outside compressors for air conditioners or heat pumps.

Draws to other subs are based on your guesstimate of how much they have done and how much there is left to do. For example, painters who are painting the interior for $1 per square foot may require some money at the end of the first week. They and all subs should be instructed to call you the day before they need a check so that you can inspect their work progress and make an evaluation of what to pay them. The more you discuss this and work it out in advance, the more comfortable you will feel when it arises. Don't be shy, you are the boss!

How to Inspect

There's an old saying in the business: "If it doesn't look right, it probably isn't." Other than structural factors most of what you will be inspecting can be judged accurately using your common sense.

There is no absolute perfection in this business, and I learned a long time ago to accept a degree of perfection — a less than perfect, but acceptable job. There are hundreds of individual processes in a construction project. Perfection in all is impossible. The structural and technical matters will (or should be) inspected by local building inspectors, and/or architects. You can, for additional peace of mind, hire an inspection engineer. But on the cosmetics — painting, trimwork, flooring, etc. — *you* will have to decide if they are acceptable. If something is flagrantly wrong, don't pay until it is done. If it is not expedient or feasible to do over, try to reach a compromise. Compromise is essential in dealing with subs. It is essential to your peace-of-mind as well as to a harmonious relationship with the subcontractor.

A "walk through" after each stage of operation as outlined in chapters 10 and 11, will enable you to make a common sense appraisal of the workmanship. You don't have to be a civil engineer to tell that doors aren't *plumb*. They won't look right and they won't operate properly. The same is true for windows, and cabinet drawers. Plumbing pipes should be secure, not rattle when you shake them. In the framing stage of additions or wall moving, check measurements to see if they are in accordance with the plans. Look, feel, examine; it's not hard. At least with an R.O.S., you don't have to worry about the correctness of the dimensions in the basic structure; it is already there. If you feel uneasy at any point, or until you gain confidence in yourself, call in one of your pros. Even plan in advance and allow consulting fees for inspections in your cost estimate.

Supervising Your Subs

Many of your subcontractors are capable of offering advice to solve problems you may encounter. For example, in determining where to run a waterpipe for a new bath, just ask your plumber. If you checked references, you should have no fear of asking advice. Subs have to stand behind their work and their advice. But don't let them offer advice for another sub's trade. Period! They are experts in their own field, not another's. I rely on my subs' advice quite often — so do most general contractors. After all we are not expert in their fields.

Most of you will be dealing with subs for the first time, other than perhaps having called a plumber or a TV repairman. Subs are, for the most part, fine, honest, and proud individuals. They are also fiercely independent. That's why they are subs. You should not, and cannot, infringe on their independence. True, you hired them, but they really don't work for you, they work for themselves. They are merely doing a job for you. That may seem like splitting hairs, but it's not. You must treat them with respect and as independent businesspeople. You don't have to, and shouldn't, watch every move they make. Besides being extremely boring for you, it is quite unnerving to them. Go over the plans and specifications with them before they start and again if needed later on, but for the most part let them do their job. Be fair but let them know you expect their best effort. You might tell them you were impressed with the job they did for so-and-so, and that you hope everything will go as smoothly on your job. Use your common sense in all communications with them and don't be afraid to ask questions.

Be sure that all your subs have supplied you with a copy of their *certificate of insurance*. A typical example is in appendix I. This certificate shows you that each sub is carrying insurance on himself and his employee, should anyone be injured on the job. Your own builders risk or fire insurance (see chapter 10) often requires that your subs carry insurance. Your own common sense, or your attorney's, should also tell you this.

Doing Your Own Labor

I have very strong opinions about people doing their own labor in building and restoring a house. It takes years for one to learn the various trades and little of that knowledge comes from books. It comes from practical experience. Most of us don't have the experience to do even one of those trades proficiently, let alone a couple of them. But more importantly, do-it-yourself labor usually takes at least twice as long. Because of today's high interest rates, this means you usually end up paying more by taking the extra time to do the job yourself than you would pay having a professional do it. Also add the facts that (a) you could hurt yourself, even seriously; (b) you could lose time from your job or family; (c) you could possibly do an inferior job, and I think you'll agree it is wise to confine yourself to the role of general contractor. That's a pretty big role by itself, but one you can do well.

If you are determined to do some of your own labor, then I suggest you limit yourself to safe and relatively easy tasks like sweeping and clean-up, landscaping, window washing (no ladders or hanging out of windows, please), and minor touch-ups. Try to keep to things that, if they do take longer or don't get finished, won't hold up completion and cost you extra interest. Don't attempt to do the jobs that building codes or your insurance company may require pros to do. Unless you are an electrician by trade, you may not be able to legally wire a house. Considering that many home fires are electrical, I think it is wise you are prevented from doing so. Of course, if you are an electrician by trade, then by all means do your own wiring. The same goes for the plumbing if you are a plumber. Codes also usually prohibit non-professional plumbing, heat-A/C installation, and other work. You will have to check with your local building inspection department. Many people think that they can do their own painting, and many do. But keep in mind that this job comes toward the end, when you need to get the project finished because your interest expense is growing. Pros come in and finish fast. Also keep in mind painting involves exterior work that can be dangerous. Professional painters have the proper ladders and scaffolding, and are used to working at heights. You could hire a painting sub for just the exterior, but he will charge more per square foot if he doesn't get the inside work also, for that is the "gravy" part of his contract.

Finally, remember that you don't have accident insurance as your subs do. Nor can you get it at a reasonable cost from any insurer, for you are not a pro. That seemingly easy roof repair job or reshingling could be a financial as well as physical catastrophe.

Chapter 9

Suppliers

Shopping for suppliers for restoration is like shopping for anything else. The only difference may be the amount of money you are going to have to spend. But, suppliers are exactly like department stores except they sell different items.

Kinds of Suppliers

The suppliers you will most likely be using are:

Building supply company. For purchasing framing lumber, nails, interior and exterior trim, windows, doors, roofing, siding, hardwood flooring, paneling, sheetrock, and quite often materials that used to be carried only by specialty suppliers: floor covering, Formica, plumbing supplies, light fixtures, appliances, paint, insulation, tile, and cabinets. The recent change in building supply companies to full-line, one-stop centers, is to your benefit; record keeping is simplified. The fewer suppliers you have to deal with, the easier it is for you to get prices and keep track of money spent. Frequently, building supply companies provide installation of many of their materials (except carpentry, plumbing, and electrical) or will have a list of qualified subs to do the job. Conceivably, you could, through one building supply company, find all your needed supplies and subs. Be sure to compare prices with other suppliers and subs. Almost all major building supply companies are listed in the Yellow Pages.

Floor covering company. Although building supply companies carry flooring and Formica, the specialty stores for these items seem to carry a wider selection and at very competitive prices. They, too, are listed in the Yellow Pages.

Lighting Fixture Company. This specialty store too, will probably have a greater selection for you to choose from than a building supply company would have.

Paint store. Your painter should, if at all possible, supply paint in the bid, but you may need to shop for wall paper. Paint stores are listed in the Yellow Pages.

Appliance store. For best selection and competitive prices on stoves, dishwashers, etc. Have the supplier install these if at all possible. That makes them responsible for the dents and scratches that electricians are famous for when they install. Again try the Yellow Pages.

Insulation supplier. I recommend a specialty supplier because they are the pros who can answer your questions regarding how much insulation is needed and the best way to install it. Insulation suppliers are listed in the Yellow Pages.

Tile company (for ceramic tile). Tile companies can sometimes match tiles from generations ago or make suggestions. They also supply and install marble, slate, and stone. Get three or four references from jobs they have done and be sure to check them. These are decorative items and the care and craftsmanship of their installer is most important. Tile companies are listed in the Yellow Pages.

Mail order houses. At your local library you can find several magazines that advertise the sale of such hard-to-find decorative items as brass fixtures, leaded glass, glass doorknobs, etc. Check the reference desk at your library. There are quite a few of these specialty suppliers, and you may be fortunate enough to have one in or near your city. Check the Yellow Pages to see.

If you are involved in a major rehabilitation project, involving foundation and structural work, you may also need the following suppliers:

Sand and gravel company. To purchase stone for drives and drainage and sand for masonry work. See the Yellow Pages.

Brick company. These suppliers carry decorative and face brick plus pavers for floors and patios. They can often match old brick or make suggestions. See the Yellow Pages.

Concrete block and brick company. For buying mortar mix and foundation materials. See the Yellow Pages.

Concrete supply company. You may need concrete for basements, garages, drives and walks. I prefer to have my concrete subcontractor (concrete finisher) supply the concrete and include it in his price.

You will probably need the first set of these suppliers. Open your accounts early so that there are no delays. Be sure you feel comfortable with the suppliers and their employees. This is most important for, as you will see, they can help you in many other ways.

In the early stages of your decision-making process, you can get price quotations over the phone even before you open an account.

How to Open an Account, Receive Discounts, Order, and Pay

To open an account with a supplier is like opening a charge-account with any department store, but easier. Three credit references and/or a bank reference are usually all that is required.

You do not have to be a licensed general contractor in order to buy at a builder's discount. Remember, you *are* a builder. You will be personally responsible for any credit secured, as are licensed builders.

You may find it expedient or necessary to contact the manager of some supply houses in order to be quoted builders' prices. This is only for their protection and should be a mere formality. It is to prevent anyone from walking in and obtaining builder's discounts for small backyard projects. But when you explain the scope of your project and they see the amount of business you may be giving them, you will have no problem.

Discounts may not be as large as you expected, as we discussed earlier in the book, but they do add up. Some cash and carry places offer no further discounts to anyone. I don't usually patronize those places since I need credit in order *not* to use my own money during a project, but to use instead the money from the construction loan which is disbursed only after portions of the work are complete.

In order not to use your own money to pay for materials or as little of your own money as possible, it is necessary to understand the terms which most suppliers give.

Most suppliers "close their books" between the twenty-fifth and the last day of each month. They then bill their customers for purchases prior to this closing date and expect payment by the tenth of the following month. If you pay by the tenth of the following month, you often earn an additional 2 percent discount. Check and see, as this is often not stated on invoices or statements. It is not a great amount of money, but it all adds up. If you time your ordering so that you purchase just after the suppliers close their books for a particular month you can have anywhere from forty to forty-six days to pay that bill (from the twenty-fifth until the end of the following month, plus ten days). A June 1 purchase would be due July 10. This gives you time to use materials, get an inspection from the savings and loan, receive a construction draw and pay the supplier's bill with construction loan funds. Obviously, you can't always time your purchases to take full use of these terms, but when you can, it helps.

This time lag also allows sufficient time to return unused materials that weren't needed and thus lower your amount due. Most suppliers allow returns of unused merchandise. Some charge a restocking fee and/or a pick-up fee. Check beforehand so you won't be surprised or angry if and when the need to return merchandise arises, and it almost always does.

Most specialty items such as special windows and plumbing fixtures cannot be returned. Again, ask first.

Other Services Suppliers Provide

In today's competitive building market, suppliers are eager to serve you. They are eager to make a sale and will offer you other services that were once reserved for their larger customers. For example, from your plans a supplier will estimate for you the materials you need. A typical materials list, a "take-off," is shown in appendix F. It was prepared by one of my suppliers. These lists are usually quite accurate and save you time and expense.

Suppliers will also schedule deliveries of materials to coincide with construction progress. In many cases, the salesperson will actually keep an eye on construction progress by going by the job site and then releasing for delivery the portion of your order that is needed at that time. Not all will do this but check. It costs nothing to ask.

Many suppliers, such as lumberyards, have computerized kitchen and bath design services.

Record Keeping

Record keeping consists of keeping bills in a paid or unpaid file and using a separate checking account to record payments. You will also want to keep a file for contracts with subs and paid bills for subs; a file for permits and fees; one for insurance forms; and one for prices (quotes) from suppliers. For one project, this easy method of keeping records should suffice.

Chapter 10

Final Planning

Building Codes

Not all states or locales have or enforce building codes. It is your responsibility to check with your local city or county government and see what if any codes prevail and how they are enforced. You do *not* have to be familiar with all the codes. It is the responsibility of each sub to comply with the codes which pertain to that trade. If there are codes, you may want to obtain a copy of the code book and become familiar with it, but you don't have to.

Codes are your safeguard that any work done on your project will enhance the safety and security of your dwelling by adhering to sound construction principles. Where there are no codes, you may want to have an inspection engineer inspect the work of the various trades. Where there are codes, inspectors who are paid by the local government will do this for you.

Building inspectors are considered a nuisance by many professional builders because they often slow down a job by insisting that some discrepancy be remedied before work progresses. I consider them my allies in that they watch out for improper or unsafe building practices for me. I certainly hope you will feel this way too. Your local building-inspection department will gladly advise you as to what inspections are performed and when to call for them, although you should not have to call them, for that should be the responsibility of each sub involved. Insist on this in your early negotiations with each sub. Remember not to pay a sub until his work is inspected and approved by the building inspector or your inspection engineer. You will be responsible for inspections relating to carpentry, insulation, and where applicable, a final building (structure) inspection.

Utilities

Most O.S.'s have existing utilities so you will have water, sewer, and electricity at the job site. But if not, or if you will be doing a total gut job, your subs can make provisions for temporary water, sewer, and electricity service. Perhaps there is a bathroom that can be gutted last. Bathroom facilities are important, and required in some areas. A portable toilet will have to be rented if no other facilities are available. If it is winter, be sure that there is heat in the R.O.S. to prevent water pipes from freezing. If not, have the pipes drained by your plumber. Temporary heat can be arranged by your heating/air-conditioning sub or your electrical sub. You may want to install new heat as one of the first phases of restoration. Work it out with your subs and let common sense prevail. Be sure any charges for temporary services are included in their contract price.

Be sure all needed temporary hook-up or permanent hook-up utilities are ordered. All can usually be arranged by phone. It will take about one hour. If repairs are needed on existing utilities to make them operable, at least during construction, then the appropriate sub may have to be called.

If you are adding or repairing a well and/or septic tank, this is the appropriate time to do so. A well takes a few days to install, a septic tank five days. Repair to either can range from one hour to one day.

Insurance

When you buy the O.S., you will want to arrange for fire insurance. Your lender will require that it be in force before any money is disbursed, and you will want to be sure it is in force *as soon* as you buy the O.S. anyway. Be sure the coverage includes the dollar amount of the restoration work to be done. A call to your insurance agent will answer any questions you might have.

Be sure your liability insurance is adequate. Construction sites can be "attractive nuisances," especially to neighborhood children.

Permits

Before any actual work can begin on your project, you will need to secure the necessary permits, if any, and pay any fees required for your locale. Call your city or county government and they will inform you of the proper procedures and which departments to visit with your plans.

The permits and fees are usually under the jurisdiction of the building-inspection division of city and county governments. Procuring the permit starts the process and gives the inspection department power to inspect for compliance with codes. The whole process of obtaining permits shouldn't take more than an hour.

Cleaning

Have all trash removed and the R.O.S. swept clean, if necessary, before you start. Have each sub keep it clean thereafter. You'll get a better job if the workmen aren't tripping over debris. Cleaning should take a half day to one day. Have a dumpster on site.

Telephone

I recommend that you have a telephone installed at the O.S. so that you and your subs can stay in touch in case additional materials are needed. It is especially important to be able to talk with your carpenter. Subs can order supplies directly if you prearrange this with the supplier, or you can order for what the subs will need. I let my carpenters order additional materials, but you may not want to. If the O.S. can't be locked up at night, ask the carpenter to take the phone home with him. I have never had my "job phone" abused.

Part III

Getting It Done

Chapter 11

Tearing Down and Building Up

You'll find that your renovation will happen in two phases: tearing down and rebuilding. Both stages have their unique set of challenges. This chapter will help you to know what to expect.

Tearing Down

Following is the general order of steps and the approximate time for the completion of the tearing down phase of a renovation project.

Wall Removal. For renovations the first step is wall removal, per your master plan and blueprints. If only a few walls are to be removed, your carpenters can usually handle first the removal of the plaster or wallboard for those walls only. If all the plaster is to be removed from all the walls, remove the plaster before tearing out any walls. Your carpenter may want to hold off on wall removal until ready to do any additional framing or bracing. This is fine. Discuss this scheduling, as well as costs, in the planning stage.

Plaster Removal. Other than cutting holes to change wiring or plumbing, I recommend removing plaster only if it is severely cracked or falling down. Patching is always cheaper — even covering cracked walls with drywall is less expensive than total gutting. Keep in mind that all interior trim must be removed, and can seldom be saved, in total gutting. Allow one-half to one day per room, including trim removal.

As mentioned in estimating, plumbing removal may necessitate removing some plaster, but usually in small areas that can easily be patched. Time will vary with the difficulty of the job and the extent of work to be done, but one to two days should suffice.

Wiring and Plumbing Removal. If you are doing a total gut and have removed all the plaster or drywall, then this is the time for removing any plumbing pipes, electrical wiring, radiator pipes, or boilers you have decided to replace. All this can generally be done in one week.

For New Additions. Clearing, grading, excavation, and hauling away trash are the first steps for new additions. Time: one to three days.

Protection from Weather and Vandals. You must protect what you already have in your O.S. from the elements and vandals. Have your carpenter install temporary locks and board-up missing windows. If the roof is in serious disrepair, it should be among the first items rebuilt (see Roofing under Rebuilding, below), but at least cover it with plastic for now. Water can do more damage than vandals.

Rebuilding

The following is the general order of steps and the approximate time for completion of the rebuilding phase of a renovation project.

Foundation, Concrete, and Brick Work. Adding or repairing footings and foundations, slab work in basements, and such brick work as chimneys, new foundations or other brick repairs can be completed before any of the items in this chapter is begun, but I prefer to tear into an O.S. and do all the removal necessary before I start repairing anything. The only exception is the roof, which I will discuss momentarily. Allow one week for repair or additions of footings and foundations.

If your inspection engineer has reported or if there is evidence that water is or has been in the basement or crawl space, now is the time to call in your waterproofing subcontractor. Both can be waterproofed from the inside, but it is best to do it from the outside. This often requires digging around the foundation and can be expensive, so be sure to get bids. A footing drain may be required and your waterproofing sub is responsible for this. Seek additional advice or opinions, if needed, from your architect, inspection engineer, or building inspector.

Rough Carpentry. By now you understand that your carpenter is one of your key people, for he is needed from the onset of the project for plan review through the initial stages of restoration. This is the time to remedy all sags. Sags should always be corrected prior to doing any new plumbing because correcting sags can crack pipes.

It is wise to have your carpenter ask the electrician, plumber, and heating/air-conditioning subs what he can do to make their jobs easier. For example, many old houses have receptacles in the baseboards. Removal of the baseboard facilitates replacement of wiring, but the baseboard removal is a job usually handled by your carpenter. The same is true with the removal of kitchen cabinets to aid the plumber. *Chases* may need to be constructed for pipes, heat vents, A/C lines, and service wires. This is accomplished by *furring out* a wall, or building a box-like run from floor to ceiling in closets, or wall-to-wall along the ceiling at an inconspicuous place. These chases or furred areas may or may not be indicated on the plans. If they are not, your subs can work it out on the job.

Carpentry rough-work will vary with the complexity and extent of restoration but allow a maximum of three weeks.

Roofing. As mentioned earlier, you should have your roof repaired as soon as possible to prevent further deterioration of the O.S. With additions, the roofing is completed early on for the same reason. Also, with the roof repaired your other subs can work inside on other phases regardless of the weather.

Your carpenter will repair any rotted or sagging roof framing and your roofing sub will then install new roofing or patch existing roofing per your specifications. The time it will take to complete roofing is from one day to two weeks, weather permitting.

Note: If there is to be chimney repair, or a new chimney, and you are putting on new roofing, have the roofer leave undone an area adjacent to the chimney. This will prevent damage to the new material while the chimney work progresses. Hold back 10 percent of the roofer's contract amount until he can come back and finish this area after the chimney is completed.

Electrical, Plumbing, Heat-A/C. All your electrical, plumbing and heat-A/C should be completed at this time, prior to insulation of side walls. Time to complete all three: two to three weeks.

Note: Don't change the pipes, unless you do not have good water pressure. You can change them later if necessary with little damage to existing walls. The same applies to electrical work. If existing electrical service is adequate for most of the house, don't remove it unless you are doing a full gut. You can add to it as added electrical demand requires. Of course if it is deemed unsafe, remove it.

Chimney. Chimneys can now be added or repaired. Old chimneys, if they are structurally sound (see your inspection engineer's report), should be relined with terra-cotta flue liners and new dampers, if this was deemed necessary when they were inspected. This is faster and less expensive than completely rebuilding. Time to repair: two days to one week. Time to build new: one to two weeks not including tearing down old chimneys, which could take one day to one week.

Exterior Siding and Trim. This phase can be completed even while your electrical, plumbing, and heating and air work is being done inside. Time to complete: one to three weeks.

Insulation. The cost of energy is high and sure to go higher, so it pays to insulate well. You not only save on energy bills in the long run, you get immediate savings because you can install a smaller heating and cooling system that will cost less. Virtually every structure can be insulated without removing the inside walls. This is accomplished by drilling small holes in exterior walls from the outside and blowing fiberglass or cellulose insulation through the hole into the wall-cavity. The hole is then plugged. The alternative is removal of the interior plaster or drywall of all exterior walls to install insulation batting. This obviously is quite expensive unless you are doing a full gut anyway. Blowing the insulation into the wall-cavities should provide adequate insulation for most climates, and a reputable firm should do the job neatly. You will not, however, have the *vapor barrier* that you would get with insulation from inside. You can compensate for this by painting all exterior walls with a vapor-barrier paint. Check local suppliers for types available in your area. Vapor barriers are important because moisture can escape through the walls, lessen the effectiveness of the insulation, and rot the framing. Before the insulation was added there was no such problem because the moisture could escape through the walls to the outside. Now it can't, so the object is to keep it inside the heated area. This also adds to your comfort, as moist air is more comfortable than dry air.

Be sure attic areas, crawl spaces, and basements are insulated as well as you can afford. Your utility companies can provide you with information on the amount of insulation needed for lowest rates and greatest savings. Time to complete: one to two weeks.

Drywall or Plaster. Now you are ready for installation or patching of plaster or drywall. Where possible, use drywall as a replacement for missing plaster or to cover cracked plaster. It is far less expensive and looks better. For your first project, try to find a drywall sub who supplies all materials and removes all trash. Plasterers always supply materials. In the winter you will need temporary heat not only to protect the plumbing, but to help speed the drywall or plastering job and to prevent cracking caused by freezing — drywall mud or filler, and plaster, have a water base. Time to complete: two to four weeks.

Prime Painting. I find it beneficial to prime the walls (apply the first coat of paint) as soon as the plaster or drywall sub is finished. Imperfections in the wall show up after priming and serious flaws can be corrected now or later. Discuss that with your drywall or plaster sub beforehand.

If there is no stained woodwork, the walls can be spray painted, which is faster and less expensive than using brushes and rollers. Prime painting by spraying can usually be completed in one day, even for a rather large R.O.S. If you have quite a bit of stained woodwork and molding, brushes and rollers would be used and stained areas must be masked to prevent their getting splattered. This would take one to two weeks and could be delayed until after any interior trim is added or repaired.

Interior Trim. Interior doors, trims, and moldings are repaired, replaced, or added at this time and new cabinets, if any, for your kitchen and baths installed. Allow one to two weeks.

Painting. You are now ready for interior as well as exterior painting, although exterior painting could be completed earlier — right after exterior trim. That should be discussed and worked out with your painting sub. It is usually easier to complete the whole job at one time. Painting inside and outside will take two to three weeks.

△ **Before:** *This home, with garage, is chosen for its location. The lot allows for significant expansion.*

△ **During:** *An addition to the back of the house extends into the lot where the garage sat.*

△ **During:** *By making the addition two stories as well, this renovation adds quite a bit of space to the building.*

△ **After:** *With a little landscaping, it is difficult to tell where the old structure ends and the addition begins.*

Final Trims. At this time you are ready to install any Formica, vinyl flooring, plumbing fixtures, electrical trims, light fixtures, and final trim for heating and air conditioning — in that order. Allow one to two weeks.

If you are finishing or refinishing hardwood floors, I recommend you wait until all the above is complete and have the floors done just prior to laying any carpeting. Allow three days for sanding, staining and sealing hardwood floors. You are now ready for carpeting and wallpaper, the last items other than cleanup. (You can reverse carpeting and cleanup.) Time: one week.

Cleanup. With final cleanup also comes trash removal, which should take only a few hours or one day, maximum. At this point you can complete any necessary landscaping.

Final Inspection and Loan Closing. When all is done, be sure your subs have called for their final inspections and that you have called for a final building inspection. Again, if you don't have a building-inspection department, for peace of mind, call in an inspection engineer before final payment to relevant subs. Final inspections will verify compliance with code and that everything works.

When all has been approved by inspectors, utility companies (if applicable), your lender, and most importantly *you*, you are ready to convert any construction financing to the permanent mortgage. This is usually arranged at the convenience of the lender, attorney, and you. It takes less than an hour.

Now you can move in and enjoy.

Appendixes

The following legal instruments are published as examples. Because of varying state laws, these should not be used by you unless such use is approved by your attorney.

Appendix A

Manager's Construction Contract

1. General

This contract dated ____________ is between __________________ (Owner) and ______________________ (Manager), and provides for supervision of renovation by Manager of a structure on Owner's property at _____________________________________, and described as ___________________________________. The project is described on drawings dated _________________ and specifications dated __________________, which documents are a part hereof.

2. Schedule

The project is to start as near as possible to ______________, with anticipated completion ____ months from starting date.

3. Contract Fee and Payment

Owner agrees to pay Manager a minimum fee of ________________ ($) for the work performed under this contract. Said fee to be paid in installments as the work progresses as follows:

a. down payment (due prior to start of work) $____________
b. framing complete (if any) $____________
c. roof repaired $____________
d. ready for drywall $____________
e. trimmed out $____________
f. final $____________

Payments billed by Manager are due in full within ten (10) days of bill mailing date.

Final payment to Manager is due in full upon completion of residence; however, Manager may bill upon "substantial completion" (see Section 11 for definition of terms) the amount of the final payment less 10 percent of the value of work yet outstanding. In such case, the amount of the fee withheld will be billed upon completion.

4. General Intent of Contract

It is intended that the Owner be, in effect, his own "General Contractor" and that the Manager provide the Owner with expert guidance and advice, and supervision and coordination of trades and material deliveries. It is agreed that Manager acts in a professional capacity and simply as agent for Owner, and that as such he shall not assume or incur any pecuniary responsibility to contractor, subcontractors, laborers or material suppliers.

Owner will contract directly with subcontractors, and obtain from them their certificates of insurance and release of liens. Similarly, owner will open accounts with material suppliers and be billed and pay directly for materials supplied. Owner shall make certain that insurance is provided to protect all parties of interest. Owner shall pay all expenses incurred in completing the project, except Manager's overhead as specifically exempted in Section 9.

In fulfilling responsibilities to the Owner, Manager shall perform at all times in a manner intended to be beneficial to the interests of the Owner.

5. Responsibilities of Manager

General

Manager shall have full responsibility for coordination of trades; ordering materials and scheduling work; correction of errors and conflicts, if any, in the work, materials, or plans; compliance with applicable codes; judgment as to the adequacy of trades' work to meet specified standards; any other function that might reasonably be expected in order to provide Owner with a single source of responsibility for supervision and coordination of work.

Specific Responsibilities

1. Submit to Owner, in a timely manner, a list of subcontractors and suppliers Manager believes competent to perform the work at competitive prices. Owner may use such recommendations or not at his option.
2. Submit to Owner a list of items requiring Owner's selection, with schedule dates for selection indicated, and recommended sources indicated.
3. Obtain in Owner's name(s) all permits required by governmental authorities.
4. Arrange for all required surveys and site engineering work.
5. Arrange for the installation of all temporary services.
6. Arrange for and supervise clearing, disposal of stumps and brush, and all excavating and grading work.
7. Develop material lists and order all materials in a timely manner, from sources designated by Owner.
8. Schedule, coordinate, and supervise the work of all subcontractors designated by Owner.
9. Review, when requested by Owner, questionable bills and recommend payment action to Owner.
10. Arrange for common labor for hand-digging, grading, and cleanup during construction and for disposal of construction waste.
11. Supervise the project through completion, as defined in Section 11.

6. Responsibilities of Owner

Owner agrees to:

1. Arrange all financing needed for project, so that sufficient funds exist to pay all bills within ten (10) days of their presentation.
2. Select subcontractors and suppliers in a timely manner so as not to delay work. Establish charge accounts and execute contracts with same, as appropriate, and inform Manager of accounts opened and of Manager's authority in using said accounts.

3. Select items requiring Owner's selection, and inform manager of selections and sources on or before date shown on selection list.
4. Inform Manager promptly of any changes desired or other matters affecting the schedules so that adjustments can be incorporated in the schedule.
5. Appoint an agent to pay for work and make decisions on Owner's behalf in cases where Owner is unavailable to do so.
6. Assume complete responsibility for any theft and vandalism of Owner's property occurring on the job. Authorize replacement/repair required in a timely manner.
7. Provide a surety bond for his lender if required.
8. Obtain release of liens documentation as required by Owner's lender.
9. Provide insurance coverage as listed in Section 12.
10. Pay promptly for all work done, materials used, and other services, and fees generated in the execution of the project, except as specifically exempted in Section 9.

7. Exclusions

The following items shown on the drawings and/or specifications are *not* included in this contract, insofar as Manager supervision responsibilities are concerned.

(List here any such exclusions.)______________________________

____

____

____

____

____

____

____

____

____

____

____

____

____

8. Extras and Changes

Manager's fee is based on supervising the project as defined in the drawings and specifications. Should additional supervisory work be required because of *extras* or *changes* occasioned by Owner, unforeseen site conditions, or governmental authorities, Manager will be paid an additional fee of 15 percent of cost of such work. Since the basic contract fee is a *minimum fee*, no downward adjustment will be made if the scope of work is reduced, unless contract is cancelled in accordance with Section 13 or 14.

9. Manager's Facilities

Manager will furnish transportation and office facilities for his own use in supervising the project at no expense to Owner. Manager shall provide general liability and worker's compensation insurance coverage for Manager's direct employees only, at no cost to Owner.

10. Use of Manager's Account

Manager may have certain "trade" accounts not available to Owner which Owner may find an advantage to utilize. If Manager is billed and pays such accounts from Manager's resources, Owner will reimburse Manager within ten (10) days of receipt of Manager's bill at cost plus 8 percent of such materials/services.

11. Project Completion

a. The project shall be deemed complete when all the terms of this contract have been fulfilled, and a Residential Use Permit has been issued.

b. The project shall be deemed "substantially complete" when a Residential Use Permit has been issued, and less than five-hundred dollars ($500) of work remains to be done.

12. Insurance

Owner shall insure that worker's compensation and general liability insurance are provided to protect all parties of interest and shall hold Manager harmless from all claims by subcontractors, suppliers and their personnel, and for personnel arranged for by manager on Owner's behalf.

Owner shall maintain fire and extended-coverage insurance sufficient to provide 100 percent coverage of project value at all stages of construction, and Manager shall be named in this policy to insure his interest in the project.

Should Owner or Manager determine that certain subcontractors, laborers, or suppliers are not adequately covered by general liability or worker's compensation insurance to protect Owner's and/or Manager's interests, Manager may as agent of Owner, cover said personnel on Manager's policies, and Owner shall reimburse manager for the premium at cost plus 10 percent.

13. Manager's Right to Terminate Contract

Should the work be stopped by any public authority for a period of thirty (30) days or more through no fault of the manager, or should work be stopped through act or neglect of Owner for ten (10) days or more, or should Owner fail to pay Manager any payment due within ten (10) days written notice to Owner, Manager may stop work and/or terminate this contract and recover from Owner payment for all work completed as a proration of the total contract sum, plus 25 percent of the fee remaining to be paid if the contract were completed, as liquidated damages.

14. Owner's Right to Terminate Contract

Should the work be stopped or wrongly prosecuted through act or neglect of Manager for ten (10) days or more, Owner may so notify Manager in writing. If work is not properly resumed within ten (10) days of such notice, Owner may terminate this contract. Upon termination, entire balance then due Manager for that percentage of work then completed, as a proration of the total contract sum, shall be due and payable and all further liabilities of Manager under this contract shall cease. Balance due to Manager shall take into account any additional cost to Owner to complete the renovation occasioned by manager.

15. Manager/Owner's Liability for Collection Expenses

Should Manager or Owner respectively be required to collect funds rightly due him through legal proceedings, Manager or Owner respectively agrees to pay all costs and reasonable attorney's fees.

16. Warranties and Service

Manager warrants that he will supervise the construction in accordance with the terms of this contract. No other warranty by manager is implied or exists.

Subcontractors normally warrant their work for one year, and some manufacturers supply yearly warranties on certain of their equipment; such warranties shall run to the Owner and the enforcement of these warranties is, in all cases, the responsibility of the Owner and not the Manager.

Manager: ______________________________ (seal)
date: _______________

Owner: ________________________________(seal)
date:_________________

Owner: ________________________________(seal)
date:_________________

Appendix B

Fixed Price Contract

See page 21 for an explanation of this contract.

Contractor: ______________________________
Owner: ________________________________ date:_________

Owner is or shall become fee-simple owner of a structure known or described as: ____.

Contractor hereby agrees to renovate the structure described above according to the plans drawn by ________________________ and the specifications herein attached.

Owner shall pay Contractor for the renovation of said structure $___________.

Prior to commencement hereunder, Owner shall secure financing for the renovation of said structure in the amount of $_____________, which loan shall be disbursed from time-to-time as renovation progresses, subject to a holdback of not more than 10 percent. Owner hereby authorizes Contractor to submit a request for draws in the name of the Owner from the savings and loan, or similar institution, up to the completed percentage of renovation and to accept said draws in partial payment thereof.

Contractor shall commence renovation as soon as feasible after closing, and shall pursue work to a scheduled completion on or before seven (7) months from commencement, except if such completion shall be delayed by unusually unfavorable weather, strikes, natural disasters, unavailability of labor or materials, or changes in the plans and specifications.

Contractor shall renovate the residence in substantial compliance with the plans and specifications and in a good workmanlike manner, and shall meet all building code requirements. Contractor shall not be responsible for failure of materials or equipment that are not Contractor's fault. Except as herein set out, Contractor shall make no representations or warranties with respect to the work to be done hereunder.

Owner shall not occupy the residence and Contractor shall hold the keys until all work has been completed and all monies due Contractor hereunder have been paid.

Owner shall not make any changes to the plans and specifications until such changes shall be evidenced in writing; the costs if any, of such changes shall be set out; and any additional costs thereof shall be paid in advance of the work being accomplished.

Contractor shall not be obligated to continue work hereunder in the event Owner shall breach any term or condition hereof, or if for any reason construction draws shall cease to be advanced upon proper request.

Any additional or special stipulations attached hereto and signed by the parties shall be and are made a part hereof.

Contractor ______________________________(seal)

Owner: _________________________________(seal)

_________________________________(seal)

Appendix C

Fixed Fee Contract

See page 29 for an explanation of this contract.

Contractor: ______________________________
Owner: ________________________________ date:____________

Owner is or shall become fee-simple owner of a structure known or described as: ____

Contractor hereby agrees to renovate the structure described above according to the plans and specifications identified as Exhibit A. Plans and specifications drawn on __________________________ by______________________________________.

Owner shall pay Contractor for the renovation of said structure, cost of construction, and a fee of $_______________. Cost of construction is estimated in Exhibit B. Each item in Exhibit B is an estimate and is not to be construed as an exact cost.

Owner shall secure or has secured financing for the renovation of said structure in the amount of $_________________, which shall be disbursed by a savings and loan or bank from time-to-time as construction progresses, subject to a holdback of no more than 10 percent. Owner hereby authorizes Contractor to submit a request for draws, in the name of the Owner under such loan, up to the completed percentage of construction and to accept said draws in partial payment hereof. In addition, it is understood that the Contractor's fee shall be paid in installments by the savings and loan or bank at the time of, and as a part of, each construction draw as a percentage of completion, so that the entire fee shall be paid at or before the final construction draw.

Contractor shall commence work as soon as feasible after closing of the construction loan and shall pursue work to a scheduled completion on or before seven (7) months from commencement, except if such completion shall be delayed by unusually unfavorable weather, strikes, natural disasters, unavailability of labor or materials, or changes in the plans or specifications.

Contractor shall renovate the residence in substantial compliance with the plans and specifications and in a good and workmanlike manner, and shall meet all building codes. Contractor shall not be responsible for failure of materials or equipment not Contractor's fault. Except as herein set out, Contractor shall make no representations or warranties with respect to the work to be done hereunder.

Owner shall not occupy the residence and Contractor shall hold the keys until all work has been completed and all monies due Contractor hereunder shall have been paid.

Owner shall not make changes to the plans or specifications until such changes shall be evidenced in writing; the costs, if any, of such changes shall be set out; and the construction lender and Contractor shall have approved such changes. Any additional costs thereof shall be paid in advance, or payment guaranteed in advance of the work being accomplished.

Contractor shall not be obligated to continue work hereunder in the event Owner shall breach any term or condition hereof, or if for any reason the construction lender shall cease making advances under the construction loan upon proper request thereof.

Any additional or special stipulations attached hereto and signed by the parties shall be and are made a part hereof.

Contractor ______________________________(seal)

Owner: ________________________________(seal)

________________________________(seal)

Appendix D

Carpentry Labor Bid

To: (Your name) ________________
Subcontractor:_________________
(Address)___________________________________
Date:_________________
Job address: ___________________________________
Owner:____
Area: Heated _________________________________ Sq. Ft.
Unheated ______________________________ Sq. Ft.
Decks _________________________________ Sq. Ft.

Charges:
Framing @$________Sq. Ft. x _________Sq. Ft. =$__________
Boxing & siding @ ________ Sq. Ft. x _________Sq. Ft. =___________
Interior trim @ ________ Sq. Ft. x _________Sq. Ft. =___________
Decks @ ________ Sq. Ft. x _________Sq. Ft. = ___________
Setting fireplace @ ________ Sq. Ft. x _________Sq. Ft. =___________
Setting cabinets @ ________ Sq. Ft. x _________Sq. Ft. =___________
Paneling @ ________ Sq. Ft. x _________Sq. Ft. =___________
Miscellaneous @ ________ Sq. Ft. x _________Sq. Ft. =___________

Total charges $__________

Signed: (Your name)______________________ Date:___________
Signed: (Subcontractor)__________________ Date:___________

Appendix E

Subcontractor's Invoice

Request Number:________________

To:__________________ Contractor: ____________________

____

____

Date:_____________ Contract number:________________

Change order number: _______________

Worker's Comp. Ins Co.:_______________

Job Name	Job No.	Description of Work	Amount

Total ===========

Work Completed in Accordance with Contract

Less retainage:_____________

____

(Contractor)

Net amount due: ===========

Appendix F

Typical Materials List

100 pieces 2" x 4" x 93" Spruce
4 pieces 2" x 4" x 10' Yellow Pine #2
8 pieces 2" x 4" x 8' Yellow Pine #2
10 pieces 2" x 10" x 10' Treated
10 lb. 16d nails
10 lb. 8d nails
15 squares 235 lb. weather seal roofing
4 rolls 15 lb. felt
50 lb. ⅞" galvanized nails

10 each 2'8" x 5'2" insulated windows (sash only)
10 each screens to match
10 each 2'8" x 5'2" snap-in grids

Appendix G

Savings and Loan Inspection Report

Inspection Report and Disbursement Schedule
Date________________________________
Loan No.________________
Borrower____
Location: Street/Box #_____ On _______ Side of ________________
between ________________________ and ______________________
in ___________________ Subdivision ___________________ County
ID by ____
Date Construction to Begin___________________________________
Contractor ______________________Loan Officer ______________

1. Start-up costs	1				
2. Rough clearing and grading	1				
3. Foundations	4				
4. Floor framing	4				
5. Wall framing	5				
6. Roof framing and sheeting	5				
7. Wall sheathing	1				
8. Roofing	2				
9. Well/water connection	2				
10. Septic tank/sewer tap	2				
11. Plumbing roughed	5				
12. Wiring roughed	3				
13. Heating-A/C ducts	2				
14. Insulation	2				

15. Chimney/flue	2				
16. Siding/brick veneer	7				
17. Door frames set	2				
18. Windows set	3				
19. Particle board/flooring	2				
20. Inside walls	6				
21. Bath tile	2				
22. Outside trim	2				
23. Gutters	1				
24. Inside trim	3				
25. Doors hung	2				
26. Plumbing fixtures	4				
27. Cabinets	3				
28. Heat plant	2				
29. Exterior painting	2				
30. Interior painting	4				
31. Built-in-appliances	2				
32. Electrical fixtures	2				
33. Carpet/floor finish	4				
34. Screens	1				
35. Drives and walks	3				
36. Cleaning	1				
37. Landscaping	1				
TOTAL	100				
DATE					
INSPECTOR					
INSPECTOR					

Appendix H

Estimate of Construction Cost

(Attach Plans, Specifications, and List of Materials)

Owner____

Location of Property ______________________________

________________Subdivision__________________

Square Footage: Heated_________ Storage___________

Garage__________ Carport___________ Porch________

Estimated Time to Complete _______________________

Contractor or Supervisor (circle one)

Name____

Address ______________________________________

Telephone: Business__________________

Residence_______________

Loan closing costs___________________$____________

Construction loan interest __________________________

Builder's risk insurance ____________________________

Demolition ______________________________________

Lot Cost — Lot No. ________Block No. _________________

Temporary services — Water $______Power $ ____________

Building plans and specifications ______________________________

Start-up costs — Permit $____ Survey $____Other $ _____________

Clear and grade lot ____

Excavation — Basement $__________Footing $ ________________

Foundation ____

Masonry materials: Brick______Block______Sand ______________

Masonry labor: Contractor_________________________________

Lumber and general building materials: Supplier________________

Carpentry labor: Contractor ___________________________________

Finish materials (Doors, Windows, Millwork) ________________

Roofing: Contractor ____

Guttering: Contractor _______________________________________

Waste disposal: City tap $_______ Septic tank $________________

Water supply: City tap $____________Well $ __________________

Plumbing: Contractor _______________________________________

Wiring: Contractor ____

Heating and air-conditioning: Contractor ______________________

Insulation: Contractor _______________________________________

Inside walls: Contractor______________________________________

Ceramic tile: Contractor _____________________________________

Cabinets: Contractor ____

Painting: Contractor ____

Interior $___________________Exterior $ ____________________

Plumbing fixtures ____

Built-in appliances: Supplier _________________________________

Electrical and lighting fixtures _____________________________

Floors: Finish $_____ Carpet $______ Linoleum $ ______________

Screens: Contractor ____

Drives and walks____

Cleaning ____

Landscaping ____

Operating costs: Tool rental $___ Fuel $___ Lossage $ __________

Contingency for price increases ____________________________

Contingency for changes _________________________________

Other ____

____

____

Total $===========

This cost estimate was prepared to the best of my knowledge and belief, and is submitted for the purpose of a loan application.

Signature_______________________________

Date____________________

Appendix I

Certificate of Insurance

Name and Address of Party to
Whom this Certificate Is Issued

Name and Address of Insured

Insurance in Force

Type of Insurance	**Policy Forms**	**Limits of Liability**			**Policy #**	**Expiration**
Worker's Compensation Employers' Liability	Standard	Statutory* $ Per Accident (Employer's Liability only) *Applies only in following state(s):				
Automobile Liability		*Bodily Injury*	*Each*	*Property Damage*		
❒ Owned only	❒ Basic	$	Person			
❒ Non-owned only	❒ Compre-hensive	$	Accident	$		
❒ Hired only	❒ Garage	$	Occurrence	$		

		Bodily Injury & Property Damage (Single Limit)				
❒ Owned, Non-owned and Hired	❒	$ Each Accident $ Each Occurrence				
General Liability		*Bodily Injury*	*Each*	*Property Damage*		
❒ Premises – O.L. & T.	❒ Schedule	$	Person			
❒ Operations – M. & C.		$	Accident	$		
❒ Elevator	❒ Compre-hensive	$	Occurrence	$		
❒ Products/Completed Operations		$	Aggreg. Prod. Comp. Optns.	$		
❒ Protective (Indepen-dent Contractors)	❒ Special Multi-peril		Aggregate Operations	$		
❒ Endorsed to cover contract between insured and ________	❒		Aggregate Protective	$		
________			Aggregate Contractual	$		
________		*Bodily Injury & Property Damage* (Single Limit)				
________ ________ ________ ________		$ Each Accident $ Each Occurrence $ Aggregate				

The policies identified above by number are in force on the date indicated below. With respect to a number entered under policy number, the type of insurance shown at its left is in force, but only with respect to such of the hazards, and under such policy forms, for which an "X" is entered, subject, however, to all the terms of the policy having reference thereto. The limits of liability for such insurance are only as shown above. This Certificate of Insurance neither affirmatively nor negatively amends, extends, or alters the coverage afforded by the policy or policies numbered in this Certificate.

In the event of reduction of coverage or cancellation of said policies, the Company will make all reasonable effort to send notice of such reduction or cancellation to the certificate holder at the address shown above.

THIS CERTIFICATE IS ISSUED AS A MATTER OF INFORMATION ONLY AND CONFERS NO RIGHTS UPON THE CERTIFICATE HOLDER

Date________________, 19___ By____________________________ Authorized Representative

Appendix J
Sample Specifications

SITE WORK

Subterranean termite control. Subterranean termite control shall be performed by a licensed pest-control operator for wood-destroying organisms, licensed by the state, who has been engaged in the business of pest control for a period of not less than five years.

Application Technique. Treatment shall not be made when the soil is excessively wet, to avoid surface flow of the toxicant from the application site. Adequate precautions shall be taken to prevent disturbance of the treatment, and human or animal contact with the treated soil.

Guarantee. Upon completion of the soil treatment, and as a condition for its final acceptance, the Contractor shall furnish to the Owner a written guarantee providing:

A. that the chemical, having at least required concentration, and the rate and method of application, applies in every respect with the standards contained herein and;
B. that the Contractor guarantees the effectiveness of the soil-treatment against termite infestation for a period of not less than ten (10) years from date of treatment. Any evidence of reinfestation within the guaranteed period will require treatment without additional expense to the Owner. The guarantee shall be in a form acceptable to the architect and shall be drawn in favor of the Owner, successor, or assignee.

CONCRETE

Requirements of the contract documents apply to all work in this division.

Concrete work. The work included in this division shall consist of the furnishing, placing, and reinforcing of all concrete and cement work necessary to erect this work such as all footings, tunnels, u-block beam lintels, pedestals, piers, floor slabs, cement floor finishes, exterior stoops, trenches and such other work shown and specified. Footings shall be constructed on level beds and shall be carried below the lowest established frost line. No footings shall be constructed on frozen soil. On sloping ground, all footings shall be carried to a sufficient depth to prevent undermining by erosion.

Minimum compressive strength. Concrete shall have a minimum compressive strength at twenty-eight (28) days of at least the design strength of the particular use, but not less than 3,000 pounds per square inch.

MASONRY

Masonry installation. All masonry shall be laid straight-level plumb and true per Structural Clay Products Institution's latest edition.

Mortar. Natural shall conform to ASTM C-91 for Type M masonry cement.

Brick. All brick below grade, not exposed to view, shall be new, whole, hard, burned, common brick, Grace MW.

Joint treatment. Tool joints.

CARPENTRY

The requirements of the contract documents apply to all work in this division.

Scope of work. Furnish and install all carpentry items and labor herein after noted and reasonably intended on the plans such as joists, sills, girders, plates, studs, bracing, sheathing and subflooring, decking, stripping, blocking, etc., and finish carpentry.

Measurements. The Contractor shall obtain, at the building, all measurements checked with details and be responsible for same. He shall establish the finished lines and coordinate the work of the various trades.

Quality of workmanship. All work that is to be performed under these specifications shall be done by mechanics and artisans skilled in their respective trades in order to produce first-class construction and installation of the work. All framing shall be set level, plumb, set to align, and well braced into place. (See drawings for sizes.)

Framing lumber. Framing, grounds, blocking, etc., shall be Southern Yellow Pine, No. 2, two inch dimension with an "F" of 1,200 p.s.i.

Standards and specifications. All standards and specifications mentioned in this section refer to the latest editions unless otherwise stated and shall be considered a part of this specification.

National design specification for stress grade lumber and its fastenings. All sections make reference to this specification for all lumber uses and fastenings.

Materials

Wall bracing: Let-in 1 x 4 or approved metal braces at all corners.
Exposed framing: Select structural No. 2 Yellow Pine, treated.
Subfloors: Store: 1 x 6 SYP laid diagonally. Apts: Existing. Lofts: ¾ T & G Plywood.
Roof sheathing: ½ inch plywood exterior grade or 1 x 6 kiln-dried sheathing boards.
Wall sheathing: ½ rigid urethane foam, R4, foil-faced both sides.
Exterior decking: 2 x 6, No. 2 Yellow Pine, treated.
Railings at exterior decks: 2 x 6, top cap; 4 x 4 intermediate posts: 2 x 2 balusters with 2 x 4 bottom rail.

Construction

Joists: Double joists for all headers and trimmers and under all partitions running parallel to the joists.

Blocking: For each runner joist, install solid block or 1 x 3 cross bridging. Two rows for stands over 12 feet and three rows for stands over 15 feet.

Blocking: Required 24 inches on-center vertically in walls to receive vertical siding.

Nails: Shall be stainless steel hot-dipped galvanized or aluminum for all exposed framing and siding. Nailing schedule as per approved edition of the Uniform Building Code.

Finish carpentry materials

Exterior siding: 1 x 6 masonite lap siding with 4½ inch exposure.

Wood louvers: Build in wood louvers where shown, in pine as specified.

Screening: Provide exterior aluminum bird screen on louvers and continuous strip of soffit.

Soffits and ceilings: Shall be ¾ inch A/C plywood, painted.

Wood fascia and trim: Unless otherwise noted, exterior mill work will be of "B" and better grade kiln dried yellow pine, clear and free from knots.

Interior door and window trim and base: Shall be white pine, painted.

Joints: Miter all joints.

Case work: By Owner.

Stairs: Apt. 2 — triple 2 x 12 stringers with ¾ inch SYP riser and 5/4 inch hardpine treads.

Ladders: Apt 1 and 2 (to lofts) — 2 x 8 fir stringers with 5/4 inch treads rabbeted into stringer.

Wainscot: Material by Owner. Install as indicated in drawings.

MOISTURE PROTECTION

Roofing and related work

Built-up roofing: 5 ply, 20 year bondable, commenced washed opaque gravel topping.

Rigid Insulation: Shall be square edged, coated top surface, noncombustible fiberboard, 2 x 4 feet with "C" value of 0.12, shall be fesco board. Contractor's option: rigid fiberglass or rigid fiberboard with minimum thickness of ½ inch and equal "C" value will be considered equal.
Gravel Stops and Flashing: 26 gauge, hot-dipped galvanized iron, painted.

Installation. All roofing or related work shall be installed as per the manufacturer's recommendations for a water tight installation including projections through the roof. Installation shall be only by roofing mechanics experienced in the roofing type to be installed.

Metal roofing. Metal roofing shall be 5-V crimp galvanized roofing. Provide ridge cap and nails.

Cleaning. Clean all tar from exposed surfaces.

Guarantee. Roofer shall furnish a five-year written guarantee covering all defects connected with the installation of the roof.

Dampproofing and waterproofing through wall flashing. Provide flashing around all openings and wall over door and window heads and under sills plus any incidental flashing necessary to make the building watertight.

Caulking. Natural color by Tremco in strict accordance with the manufacturer's recommendations.

Roof accessories
Roof vent: Painted galvanized iron.
Insulation: Building paper shall be 15 pounds asphalt saturated felt.
Wall insulation: Shall be fiberglass wool batts, thickness to fill full width between studs, R13.
Floor insulation: Shall be fiberglass wool batts, thickness to fill full width between joists, R19.

Roof insulation: Rigid at built-up roof, 6 inch fiberglass batts at rafters.

DOORS, WINDOWS AND GLASS

Requirements of the Contract Documents apply to all work in this division.

Finish Hardware. Quality of hardware will be Builder's Hardware manufactured by Kwikset or equal and shall be approved by the architect prior to use.

Wood doors

Entry doors: Shall be solid core, 1¾ inch thick, 3 feet x 6 feet 8 inches.
Bi-folding doors: 1¾ inch wood paneled.
Interior doors: Solid wood paneled 1¾ inches thick.

Windows. Existing windows shall be reworked to original condition. Care shall be taken to salvage existing glass for reuse.

Storefront. Sizes on drawings are approximate. Actual sizes shall be determined by measuring in the field. Glazier shall check all field dimensions prior to fabrication and delivery.

Sheet glass. Shall be new and clean; sizes as noted or required by the Uniform Building Code.

Insulating glass. ½ inch thick, clear.

Tempered insulating glass. ½ inch clear; use at entrances, glass filled doors and storefront.

Glazing compound. Shall be permanently elastic.

Cleaning. Clean all glass surfaces after installation.

Weatherstripping. Provide bronze interlocking type weatherstripping at head and jamb of all exterior doors.

Window Weatherstripping. Shall be integral factory installed.

Threshold. Provide oak saddle type threshold at all exterior doors.

Finish hardware

Schedule: Hardware schedule shall be submitted to the architect for approval.

Lock sets and door pulls: Kwikset.

FINISHES

Gypsum Wallboard. Wallboard shall be ½ inch gypsum sheetrock or ⅝ inch fire rated sheetrock as noted on the plans.

Edging. Provide metal edging at exposed edges and external corners.

Nailing. Deep set nails or use screws to prevent popping. Compound each nailing point and sand smooth.

Joint treatment. Taping compound joints per manufacturer's recommendations; sand after each application of joint compound.

Holes. All holes cut for switches, receptacles, light fixtures and plumbing pipes are to be made with drywall cutting tool so that all cover plates completely conceal all rough edges.

Finish. Painted flat by Owner.

Painting. By Owner.

Carpeting. By Owner.

Wood floors. Repair hardwood floors and prepare for refinishing by Owner.

SPECIALTIES

Toilet and Bath Accessories. By Owner.

Foundation vent. Aluminum screen on treated 1 by 2 frame nailed to interior face of foundation wall.

Soffit screen vent. Continuous aluminum insect screen.

Equipment
Prefinished cabinet work and appliances by Owner.

MECHANICAL
Heating and air conditioning. Installation as per applicable codes and enforced with standards of the American Society of Heating, Refrigeration and Air Conditioning Engineers (ASHRAE).

Scope of Services. The Contractor for work provided by this section shall furnish all labor, equipment, appliances, etc., in connection with the complete installation, ready for use, of the items specified herein in strict accordance with the specifications and general conditions and the heating and air-conditioning work indicated on drawings. The Contractor shall coordinate his work with the other trades to avoid interference. All work and related items necessary to complete the work, whether shown on the drawings or specified or both, are part of this contract.

Installation. As per applicable codes and in accordance with standards at the American Society of Heating, Refrigeration and Air Conditioning Engineers.

Flashing. Provide flashing and counterflashing as necessary for a watertight job. Coordinate with roofing contractor where pipes and duct pass through walls or roof.

Duct work. Fabricated or galvanized sheetmetal, 24 gauge minimum.

Duct insulation. Provide 1 inch or 1½ inch fiberglass on galvanized metal ducts.

Vertical return duct. Shall be ½ inch acoustic lined.

Heating and cooling units. Apartment number 1 and 2; one each, 1½ ton heat-pump and air-handler by General Electric.

Store. One 3 ton heat-pump and air-handler by General Electric.

Work included. The work specified under this section of the specifications includes but is not limited to the following:

a. Electric heat-pump, air-handling and condensing units for each conditioned space.
b. Supply exhaust and return airduct work
c. Exhaust fans for bathrooms
d. Grilles and registers
e. Drain and refrigeration piping
f. Insulation
g. Controls
h. All other accessories and incidental items not itemized above but specified hereinafter in this section of work
i. Provisions for all contingencies, and supply of all labor, materials, temporary electrical service, scaffolding, fixtures, tools, transportation, etc., necessary for proper installation of all work described in this section of the specifications and indicated on the drawings

Guarantee. The Contractor shall guarantee all materials, equipment and workmanship for a period of 12 months after the date of final acceptance of the building by the architect and Owner. All guarantee failures shall be corrected or replaced by the Contractor as soon as possible after notification of such failure. Compressors shall be guaranteed for five years.

Refrigerant piping. Type ACR deoxidized and sealed refrigerant copper tubing with wrought copper fittings. It shall conform to Federal Specifications WW-T-799 and ASTM B-83-33 and shall be free from scale and dirt. Precharged tubing also factory.

Drain-pan piping. Condensate lines shall be PVC.

Duct work. Galvanized iron or steel sheets, conforming to latest edition of the ASHRAE guide in every respect, including duct weight seams, joints, construction gauges, reinforcing and assembly. Mount with heavy straphangers bent at least 2 inches under bottom edges of the ducts. Sheet metal screws shall be stainless steel, cadmium plated or zinc plated. Any duct work that vibrates, buckles, wraps, sags, rumbles, or is not airtight for the service required, shall be corrected or replaced at the engineer's discretion. Air turning veins of an approved type shall be provided in all square or short-rating elbows and where indicated on the drawings and details.

Insulation. No covering shall be applied until the work has been thoroughly cleaned and tested for tightness. The covering shall be applied in an approved manner and stripped in accordance with manufacturer's guide specifications and in no case shall be covered up or furred-in until inspected by the architect. All coverings shall be done by an approved insulation contractor qualified in this line of work. All coverings shall present a neat, smooth and finished appearance and shall be of fiberous glass unless otherwise specified.

Liquid refrigerant lines. No insulation required.

Suction refrigerant lines. Factory insulated or ¾ inch Armstrong foam plastic flexible insulation, Armorflex 22.

Cleanup. This Contractor shall keep the premises free of debris and unusable materials resulting from his work, and as work progresses, or upon request, the General Contractor shall remove such debris and materials from the Owner's property and leave all floors broom-clean in areas affected by his work. All fixtures to be cleaned to the satisfaction of the architect and the Owner.

Control system. Controls shall be provided for systems where indicated on the drawings.

PLUMBING

Scope of work. The Contractor for work covered by this section shall furnish all labor, materials, equipment, appliances, etc., in connection with the complete installation, ready for use, of the items specified herein in strict accordance with this section of the specifications and the general conditions in the plumbing work as indicated on the drawings. The Contractor shall coordinate his work with the other trades to avoid interference. All work and related items necessary to complete the work, whether shown on the drawings or specified or other are a part of this contract.

Work included. The work specified under this section of the specifications includes, but is not limited to, the following:

1. The installation of proper water services within the building.
2. Furnish and install all plumbing fixtures where shown on the drawings.
3. The installation of the soil waste vent and drain piping, connections to sanitary sewer system.
4. Furnishing complete installation of the domestic hot water heater where shown on the drawings.
5. The complete installation of all hot and cold water piping.

Rules and regulations

1. All work shall be in accordance with rules and regulations of the state applicable, local rules and regulations, and of any authorities having jurisdiction insofar as such regulations apply to the work or material provided under this contract.
2. The Contractor shall make all necessary arrangements and pay for all necessary fees, charges, and permits required for the complete installation of his work.

Materials. Materials shall conform to the designated standards of the American Standards Association (ASA), American Water Works Association (AWWA), Commercial Standards (CS), and Cast-Iron Soil Pipe Institute (CISPI).

Protection. Protect all work from damage and properly close all pipes with test plugs, screw-caps, etc., to prevent foreign matter from entering the pipes during construction.

Flashings. For gable or sloping roofs, flashing shall be standard 2½ pound all lead flashing unit with lead band at the top. Lead shall be crimped watertight around pipe in a form flashing under metal roof and built-up roof.

Domestic hot water heater. Heater shall be furnished and installed by the plumber contractor for the living unit. Heaters are to be UL approved, quick recovery, and glass lined.

Cutting and patching. In areas where it is necessary to cut concrete floors, walls, and ceilings to install soil waste vent or hot and cold water piping, the plumbing contractor shall do the cutting and patching. Piping shall be installed without critical damage to structural members. Holes drilled shall be at the center line of the structural member.

Plumbing fixtures

1. Reuse existing clawfoot tub, new fittings.
2. New water-closet and tank set.
3. Install Owner's pedestal lavatory, new fittings.
4. Furnish new stainless steel kitchen sink, twin compartment.
5. Furnish one 40 gallon water-heater with pan.
6. Connect Owner's garbage disposal.
7. Connect Owner's dishwasher.
8. Furnish hookup or ice maker.
9. Reuse existing fixtures in bathrooms.

10. Install one wall-mounted lavatory.
11. Install one 30 gallon water-heater.
12. Install new fittings for lavatory and tank sets.

Acceptance. All work furnished under this section of the specifications shall be thoroughly cleaned and ready for use of the Owner. Upon completion of the entire system covered by these specifications, a certificate of approval from the different city departments that have jurisdiction shall be obtained and then delivered to the Owner.

Guarantee. The Contractor shall guarantee and service all workmanship and materials, and shall repair or replace, at no additional cost, any part thereof which may become defective within the period of 12 months after the date of final acceptance. Ordinary wear-and-tear accepted.

Cleanup. This Contractor shall keep the premises free from debris and unusable materials resulting from this work and as work progresses, or upon request from the General Contractor he shall remove such debris and materials from the Owner's property and leave all floors broom-clean in areas affected by his work.

ELECTRICAL

Scope of Work. The Contractor for this section shall provide all labor, materials, equipment and services necessary for, and reasonably incidental to, the completion of all work shown on the drawings and detail sheets as herein specified.
Feeders, panel boards, circuit wiring, outlets and connections complete to meter box as required.
Connection of equipment as specified.
Lighting fixtures complete.
TV outlet box.
Telephone outlet boxes.

Regulations. The installation of the electrical wiring shall confirm with the National Electrical Codes, the local code, and the requirements of the local power company. All materials shall be new and shall be listed by Underwriters Laboratory, Inc., as conforming to its standards in every case for which such a standard has been established for the particular material in question. The Contractor shall effectively protect, at his expense, the electrical installation from injury during the construction period, all openings into any part of the conduit system, as well as associated fixtures, equipment, etc., but before and after being set in place must be securely covered or otherwise protected to prevent obstruction of the tools and materials by grit, or any other foreign matter. The Contractor will be held responsible for all damage so done until the work is fully and finally accepted. Conduit ends shall be covered with capped bushings.

Certificate of inspection. All interior electrical work is to be inspected and approved by the local electrical inspector before the system is energized. Duplicate certificates of this approval shall be delivered to the Owner. All fees for the above and for any other inspection and approval service required shall be borne by the Contractor.

Character of materials and equipment. All materials and equipment, except here and otherwise specified, shall be new and conform with standards specified herein. Equipment is herein defined to include conduits, cable, wiring, materials and devices, panel boards, etc. All equipment offered under these specifications shall be limited to products produced and recommended for service ratings in accordance with manufacturer's catalogue, engineering data, or other comprehensive literature that may be available to the public. Equipment shall be installed in strict accordance with the manufacturer's instructions for type capacity and suitability of each piece of equipment used. This Contractor shall obtain his instructions which shall be considered a part of these specifications.

Field Measurements. The Contractor shall take all field measurements necessary for this work and shall assume responsibility for their accuracy.

Drawings and specifications. The drawings are intended to show the general arrangement of outlets. Door swing shall be checked for final arrangement. The Contractor shall check all structural and mechanical plans and specifications so that he may coordinate his work with these trades. All outlets shall be located uniformly with respective beams, partitions, ducts, openings, etc., and the general location shall be checked before installing. Should there be any interference between the electrical outlet and other trades, the Contractor shall notify the Owner so that proper location may be decided upon. No outlets shall be installed in inaccessible places.

Permits. The Electrical Contractor shall obtain all permits required for his work, including the cost of same in his estimate.

TV system. The Contractor shall install an outlet box in each apartment. The cable to the outlet is to be furnished and installed by the Owner.

Receptacles and switches.

Receptacles: General receptacles shall be Arrow-Hart no. 5242 or Leviton no. 5014.

Receptacles for ranges: Leviton no. 5050.

Bathroom ground fault receptacles: Square D No. GFR-115, 15 amp capacity.

Weatherproof Receptacles: to be ground fault receptacles.

Switches: Boxes of a class to satisfy the conditions for each outlet shall be used in concealed work. Boxes shall be installed in a rigid and satisfactory manner. Switches shall be Arrow-Hart no. 1101 or Leviton no. 5501.

Connection to mechanical equipment. The Contractor is cautioned to note carefully other sections of these specifications describing equipment to be furnished under these sections, in order that he may fully understand the wiring requirements. All power wiring and switches for apartment air-conditioning units, exhaust fans and water heater shall be furnished and installed by the Electrical Contractor. Exposed connections to water-heater shall be made with junction boxes on the wall.

Telephone outlets. The Contractor shall make all necessary arrangements with the telephone company for the installation of telephone cables and terminals and coordinate the work to insure the installation at the proper time.

Lighting fixtures. The Contractor shall furnish and install lighting fixtures as scheduled on the drawings. All fixtures shall bear the Underwriters label.

Smoke detectors. The Contractor shall furnish and install smoke detectors.

Guarantee. The Contractor shall leave the entire electrical system, installed under this contract, in proper working order and shall, without charge, replace any working materials which develop defects, except ordinary wear-and-tear, within 12 months from the date of final inspection and acceptance.

Cleanup. This Contractor shall keep the premises free of debris and unusual materials resulting from his work, and as his work progresses, or upon request by the General Contractor, he shall remove such debris and materials from the Owner's property and leave all floors broom-clean in areas affected by his work.

Service. Provide 200 amp main service. Wire heat pump high and low voltage. Provide wiring for heat pumps, air handlers, dishwasher, disposal, stove, water-heater, kitchen circuit, lighting and receptacle circuits as per plans.

Allowances. Total light fixture allowance $600. Total equipment and appliances allowance $1200, which includes stove, refrigerator, range hood and bath fan.

Appendix K

Sample Blueprints

(Courtesy, John M. Knight)

Knight Residence, Plans

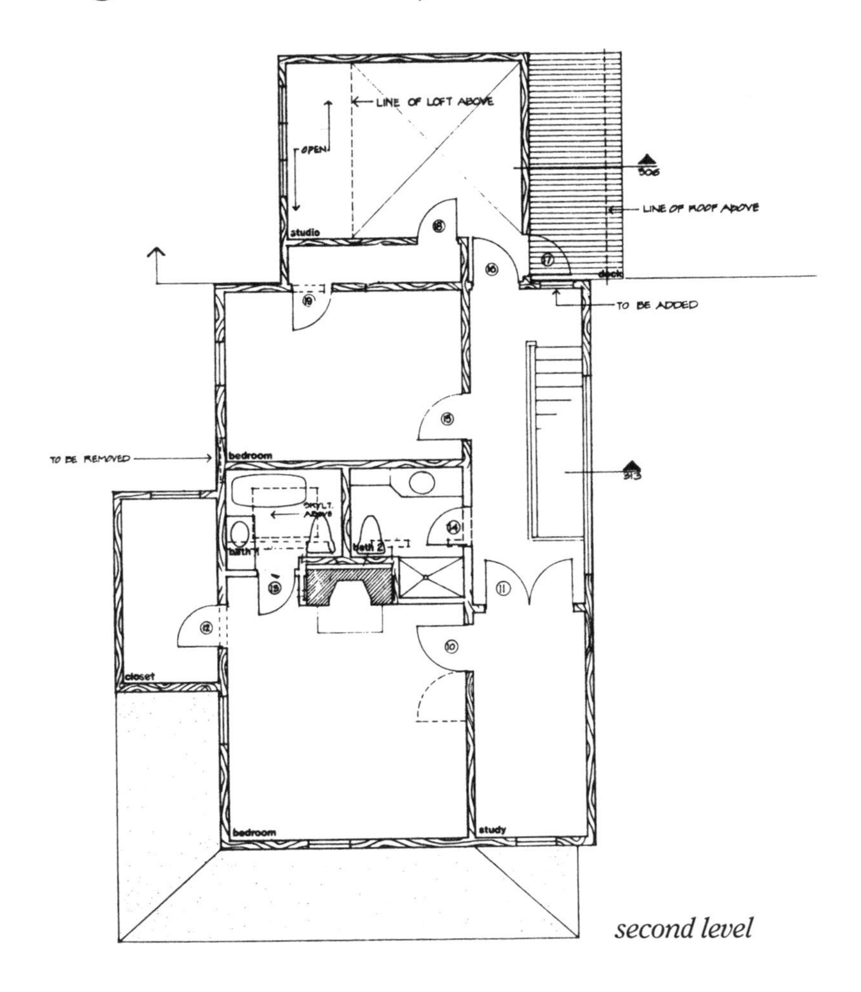

second level

DOOR SCHEDULE					
NO.	TYPE	R.O. W	H	HD. HT.	REMARKS
01					BY OWNER
02					"
03					"
04					"
05					"
06					BY OWNER
07					"
08					"
09					"
10					BY OWNER
11					"
12					"
13					"
14					"
15					"
16					"
17					BY OWNER
18					"
19					"
20					"
21					"

WINDOW SCHEDULE					
NO.	TYPE	R.O. W	H	HD. HT.	REMARKS
01					BY OWNER
02					"
03					"
04					"
05					"
06					"
07					"
08					BY OWNER
09					"
10					"
11					"
12					"
13					BY OWNER
14					"
15					"
16					"
17					BY OWNER
18					"

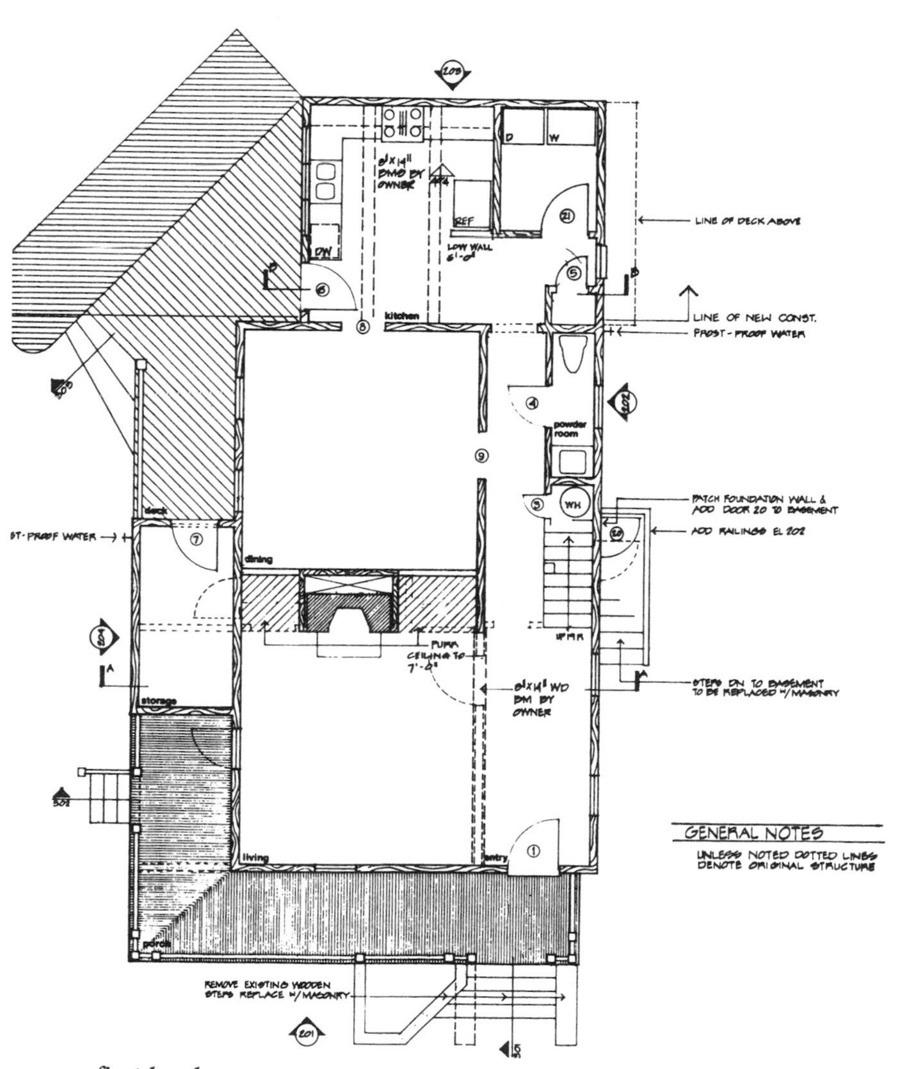

first level

Knight Residence, Elevations

north

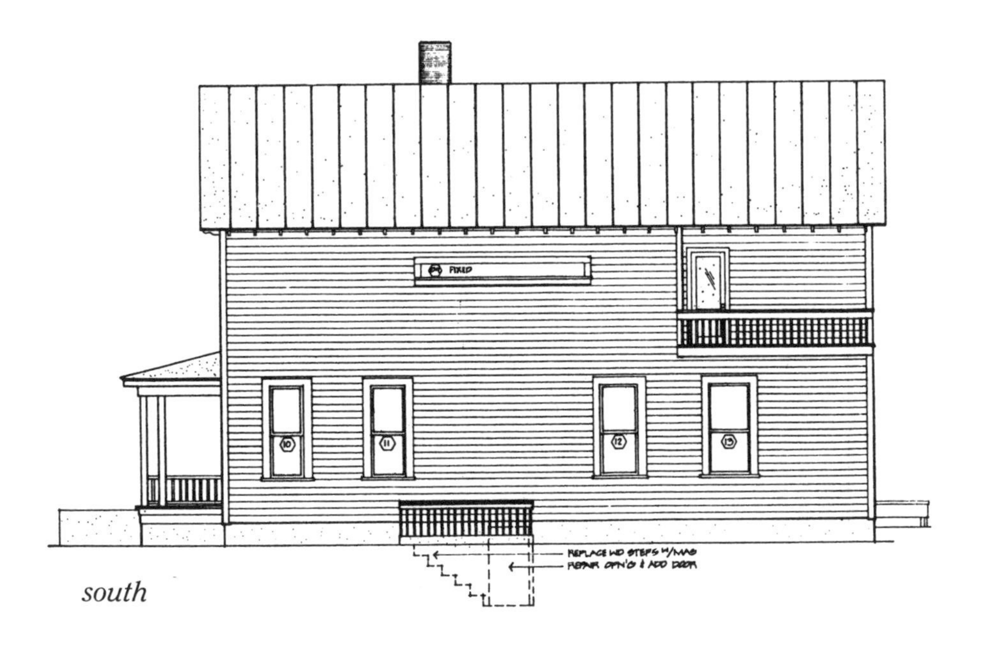

south

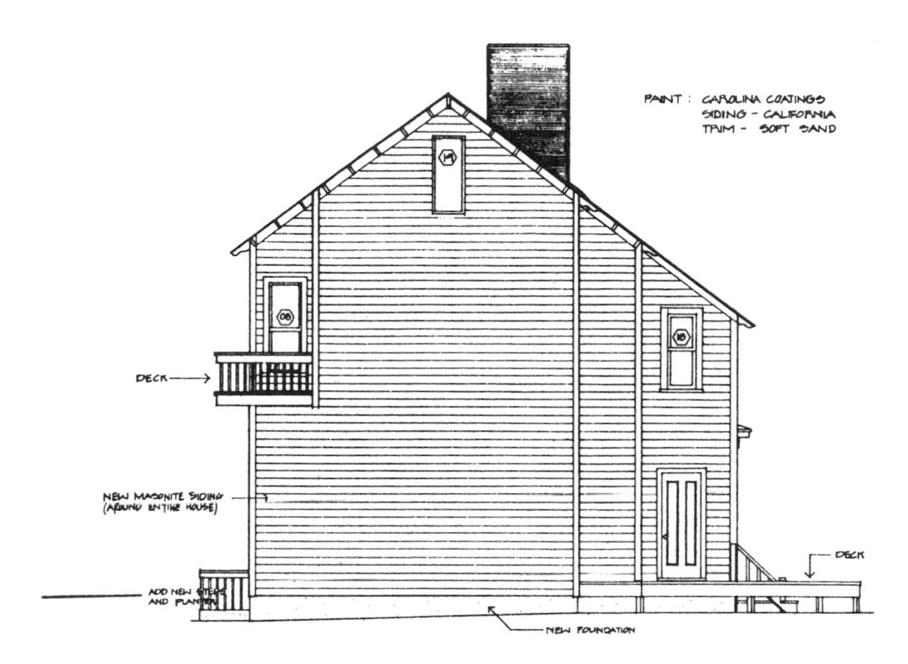
PAINT: CAROLINA COATINGS
SIDING - CALIFORNIA
TRIM - SOFT SAND
DECK
NEW MASONITE SIDING
(AROUND ENTIRE HOUSE)
ADD NEW STEPS
AND PLANTER
DECK
NEW FOUNDATION

east

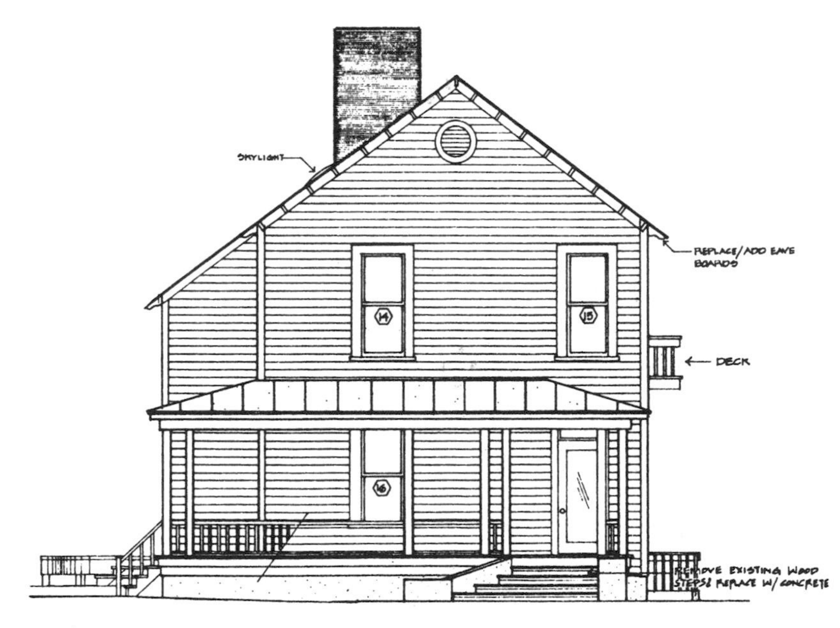
SKYLIGHT
REPLACE/ADD EAVE
BOARDS
DECK
REMOVE EXISTING WOOD
STEPS & REPLACE W/ CONCRETE

west

Knight Residence, Details

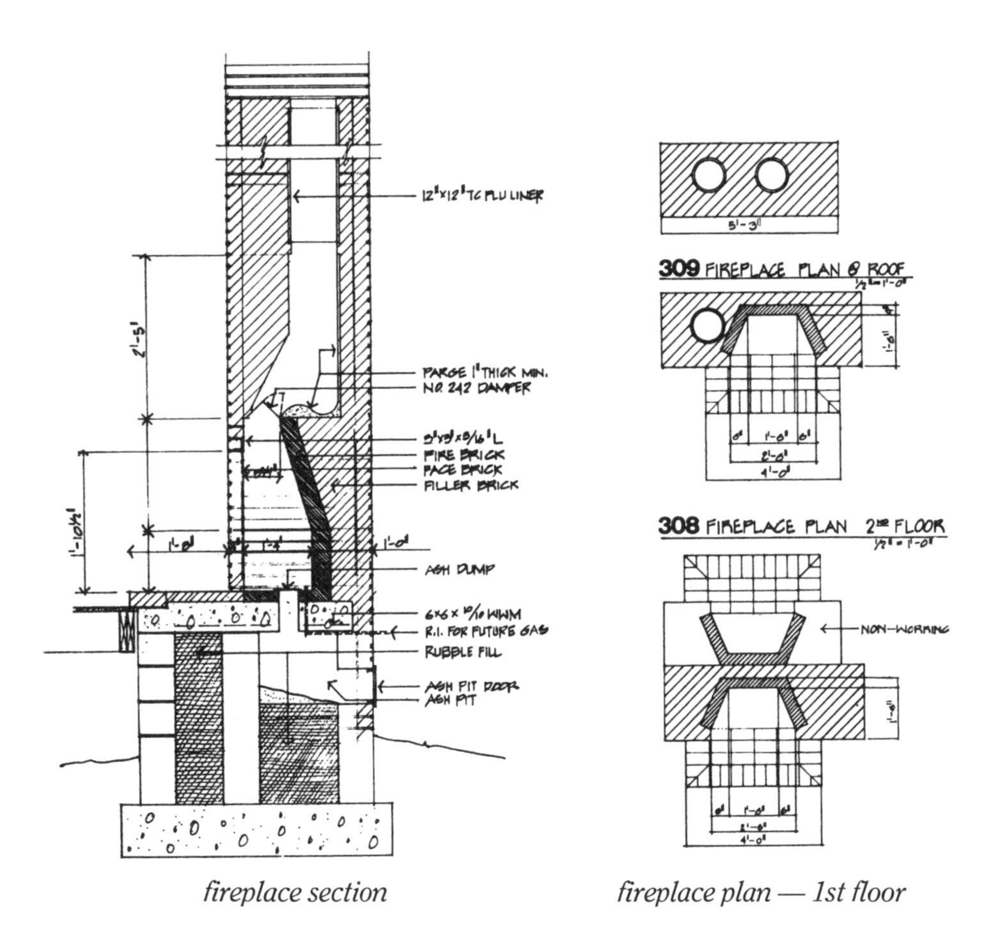

fireplace section

fireplace plan — 1st floor

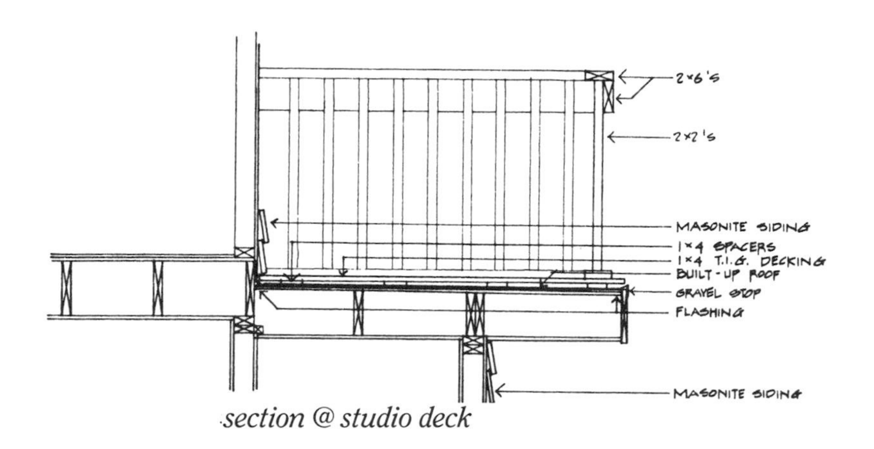

section @ studio deck

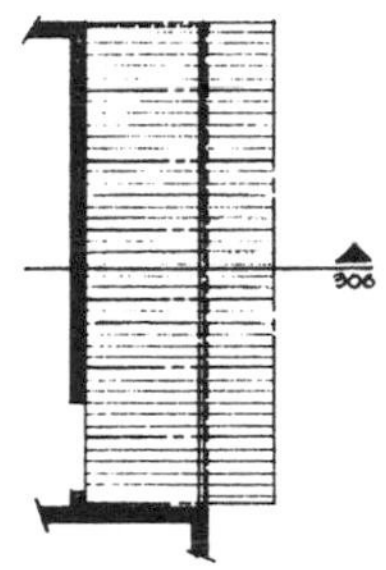

studio deck framing plan

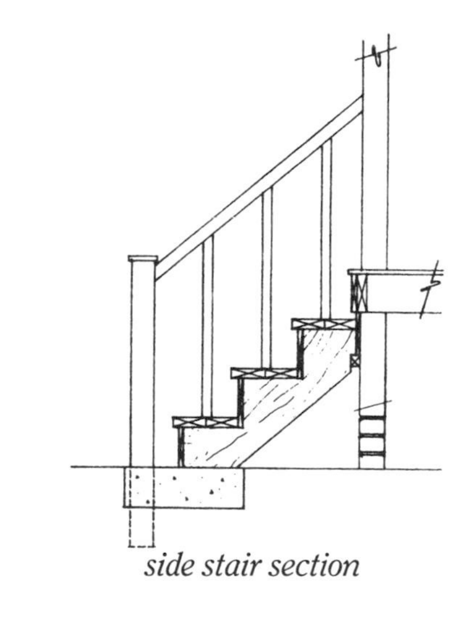

side stair section

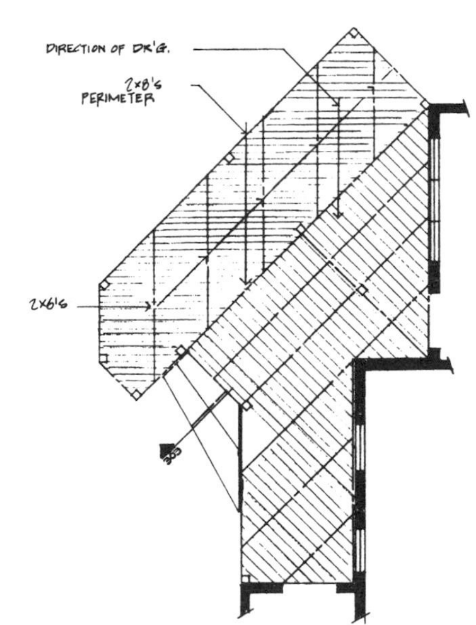

kitchen deck framing plan

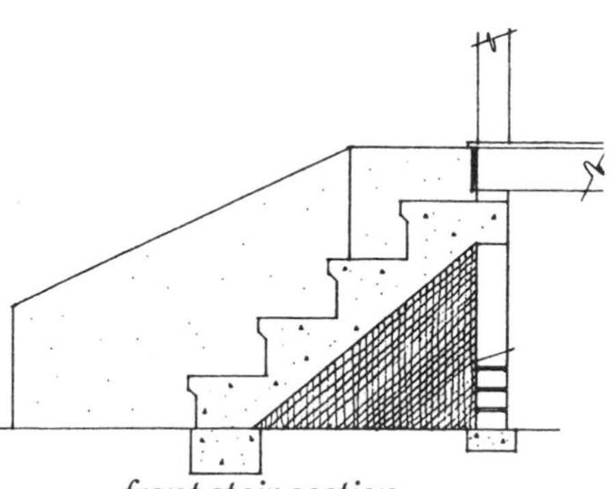

front stair section

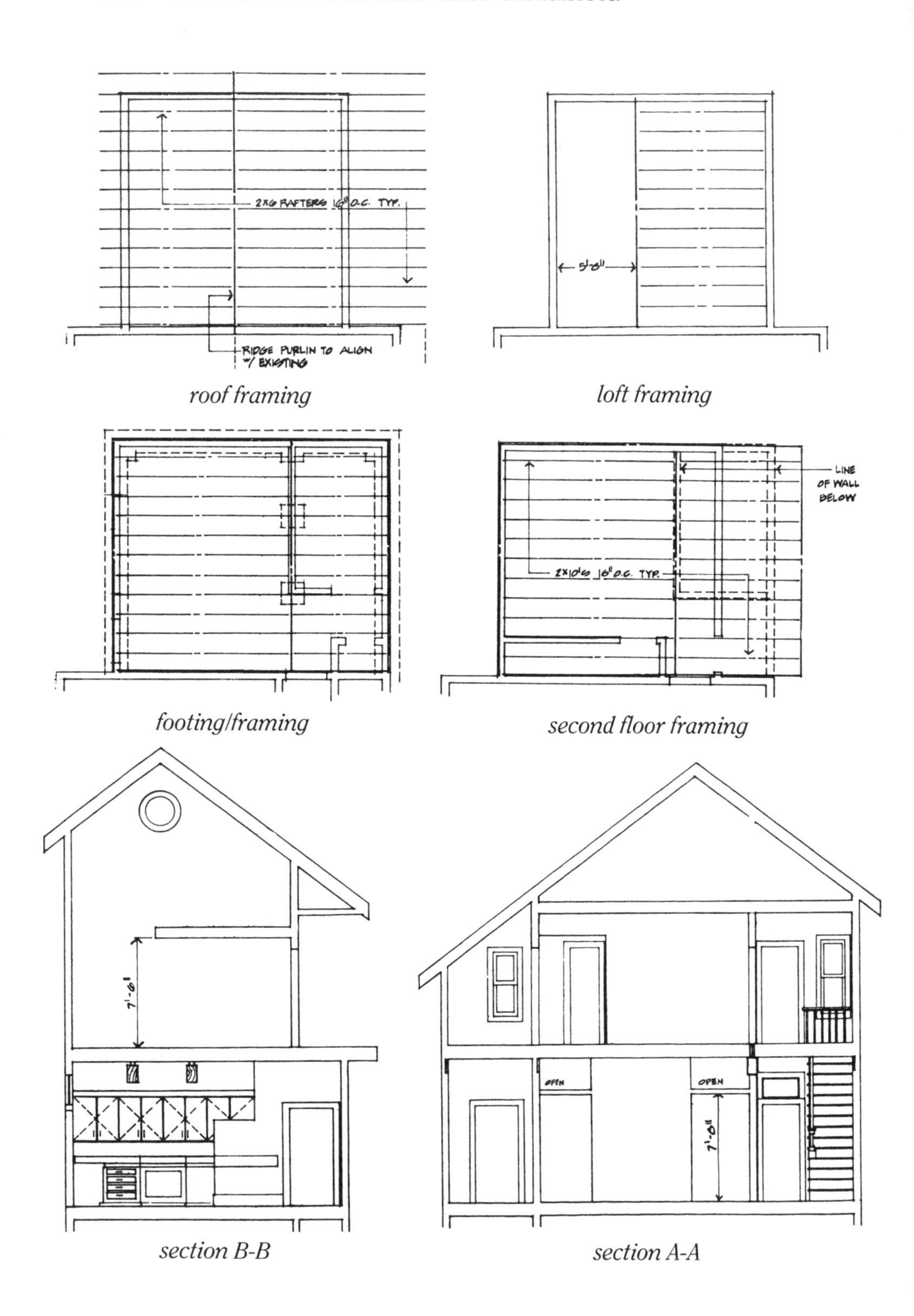

roof framing

loft framing

footing/framing

second floor framing

section B-B

section A-A

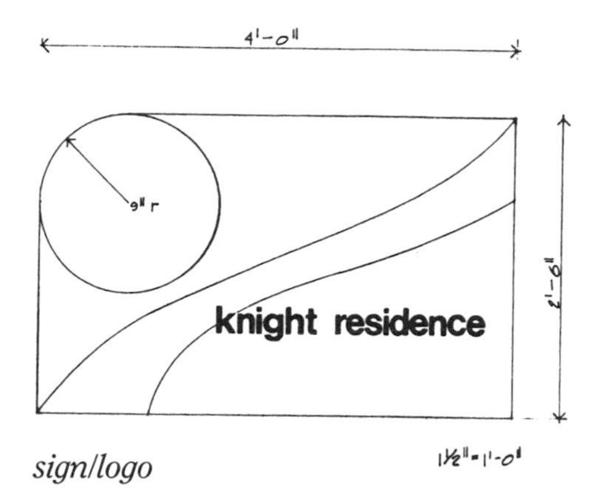

sign/logo

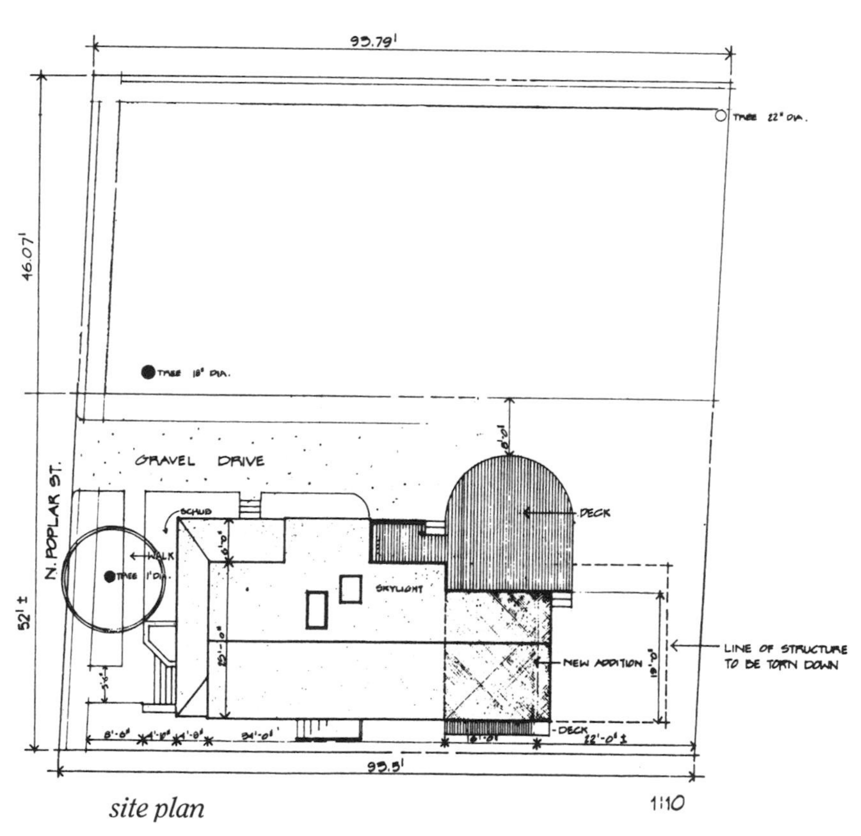

site plan

Knight Residence, Electrical Plan

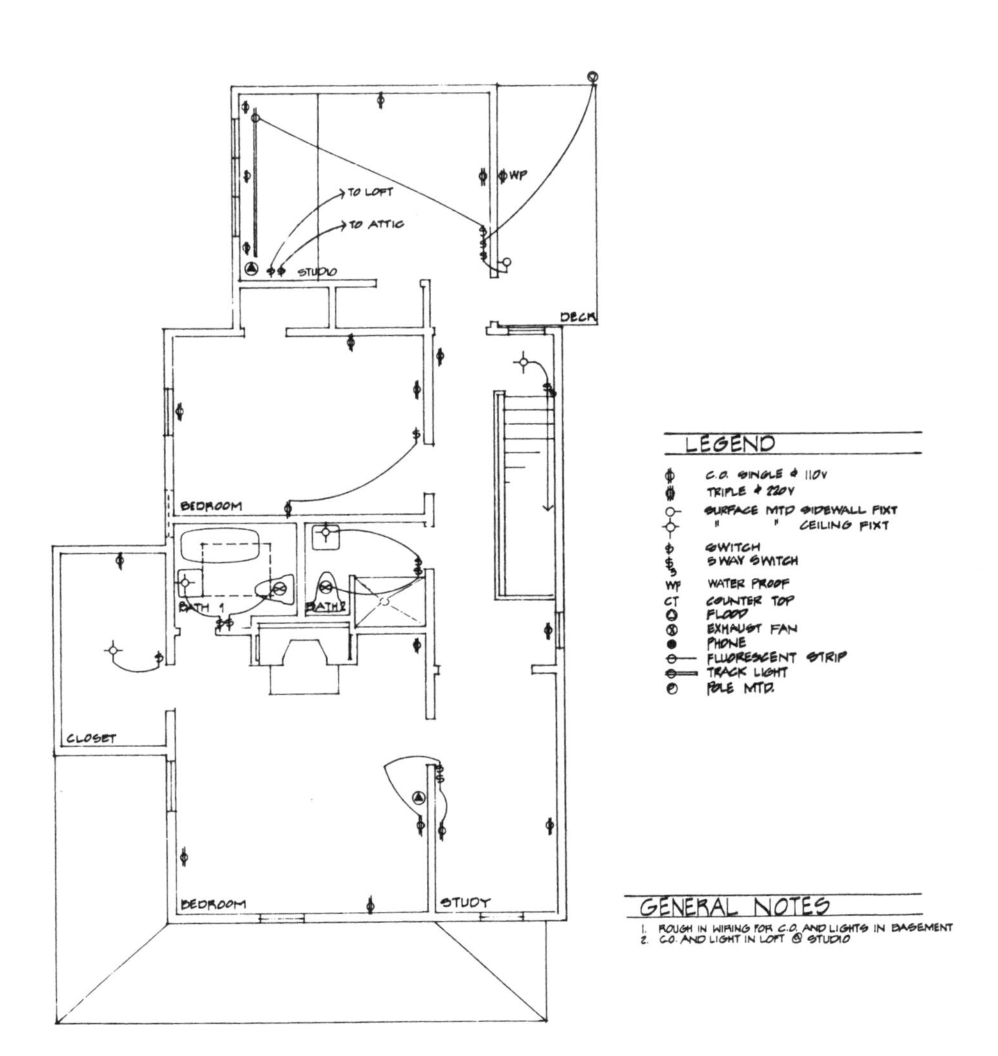

second level

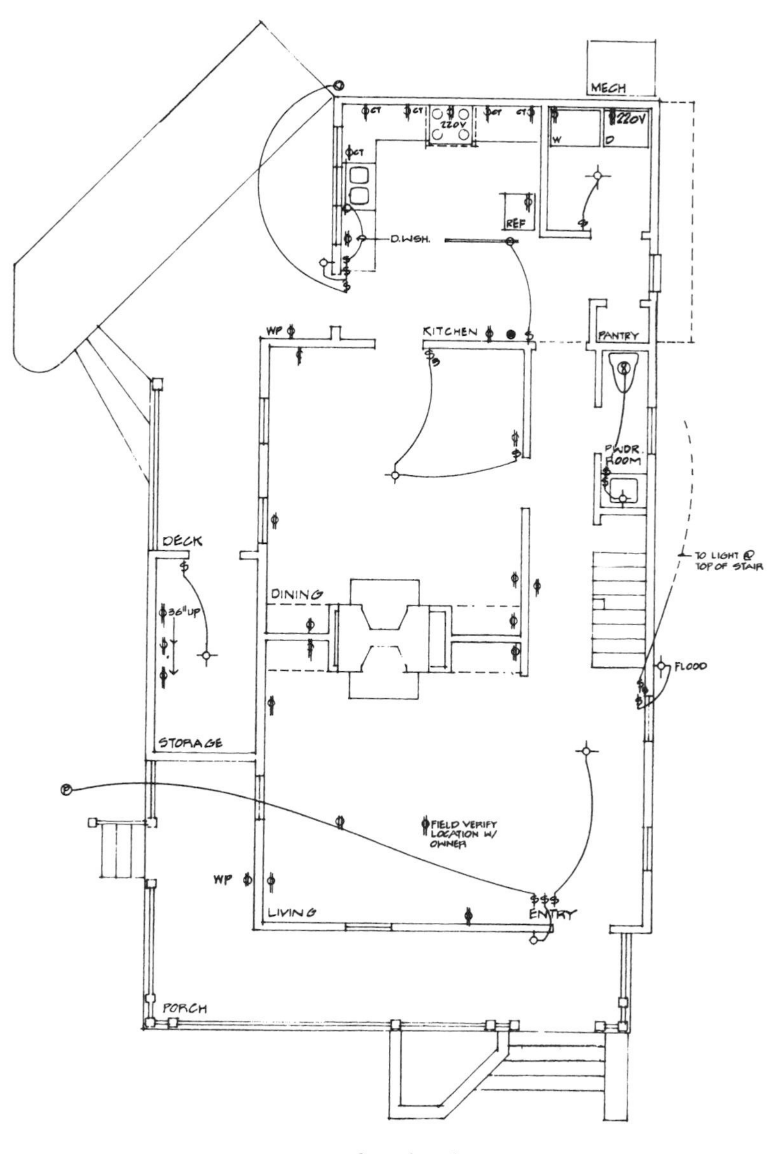
MECH
220V
CT
D.WSH.
REF
W
D
220V
KITCHEN
PANTRY
WP
PWDR. ROOM
TO LIGHT @ TOP OF STAIR
DECK
36" UP
DINING
FLOOD
STORAGE
FIELD VERIFY LOCATION W/ OWNER
WP
LIVING
ENTRY
PORCH

first level

Glossary

Appreciates (appreciation). Refers to the increase in value of a structure over a period of time.

Blueprints. A detailed plan of a structure.

Building plans. See Blueprints.

Built up roof. A type of roofing which consists of layers of felt building paper, asphalt, and stone.

Caulking. To fill cracks with a filler.

Certificate of insurance. Proof of insurance.

Chase. A channel, as in a wall or ceiling, for something to lie in or pass through — electrical wiring, plumbing pipes, or vents.

Circuit breakers. Modern safety devices that prevent electrical overloads. They can be reset after reducing the load which caused them to trip.

Circuit panels. See Panels.

Closing. Specifically, a loan closing: completion of all legal documents necessary to procure a loan.

Collateral. Something of monetary worth pledged to a lender for the purpose of securing a loan.

Compressor. Apparatus used to compress a gas called freon for the purpose of cooling, or in the case of a heat pump for heating a structure.

Contingencies. Unforeseen, or chance, additional costs.

Crawl space. An area under a floor or roof in which there is insufficient height to stand.

Deed restrictions. Encumbrances placed on a piece of real estate and filed at the county courthouse as a matter of record which restrict or prevent certain uses of that real estate.

Draw. A disbursement of money that represents only a portion or percentage of the entire amount due the person.

Ductwork. Pipes used to distribute heated or cooled air in a structure.

Durawall. Brand name of steel webbing lace in masonry walls to strengthen those walls.

Elevations. Drawings of the exterior of a structure.

Equity. The difference between what you owe on something and what it is worth.

Fascia boards. The flat horizontal boards that form a band around a roof's edge.

Fire insurance. An insurance policy that protects you financially in the event of fire damage to the structure.

Fixed price contract. A contractual agreement with a *general contractor* whereby, for an agreed-upon price, he will renovate, restore, or remodel an old structure.

Framing. All the structural members of a building.

French drains. Pipes placed around or under a structure to provide positive drainage of water.

Furnace. Apparatus used to heat air. It can operate on electricity, gas or oil.

Furr out. Making a wall or ceiling deeper in order to provide a *chase.*

Fuses. Older safety devices that prevent electrical overloads, largely replaced by *circuit breakers.* They cannot be reused after they perform their function.

General contractor. The person who manages the time, money, and people involved in a construction project.

GEM. Growth Equity Mortgage: a mortgage in which the lender shares the expected appreciation and increased *equity* of a structure. For this the lender will give a "below market" rate.

Gut. To remove all the interior of a structure, except the framing.

Heat pump. A heating system whereby heat is extracted from the outside air — even cold air — and used to heat a structure. Acts as air-conditioning in the summer.

Insulated glass. Usually refers to at least two pieces of glass with an air space in between to provide insulation.

Joists. Horizontal framing members that comprise the floor or ceiling of a structure.

Lease option. Leasing a structure with the option to buy later at a predetermined price.

Listings. Pieces of real estate for sale.

Panels. Specifically, electric panel boxes. Metal boxes that contain the *fuses* or *circuit breakers*.

Plates. Other horizontal framing members that support walls, ceiling *joists*, or rafters. Includes *sill* plate.

Qualified buyers. Potential buyers who have already been deemed financially able to purchase a particular structure.

Radon. A naturally occurring radioactive gas that emanates from the soil and is the second leading cause of lung cancer in the United States. Can be detected by testing devices available at retail stores.

Re bars. Steel rods placed in concrete to add strength.

Recording fees. Charges by a local courthouse to record and file a deed or other legal document.

Rolled over. Converted to.

Sill (sill plate). A horizontal framing member, immediately adjacent to the foundation, that supports a wall or floor.

Soffit. The underside panel of a roof overhang or cornice.

Soffit vents. Air vents in the *soffit* to provide adequate ventilation to the framing members of the roof. Prevents rot and heat build-up.

Specifications. A listing of the particulars — size, quality, etc. — to be done in the renovation of the structure.

Thermostat. A device that, based on temperature, regulates the heating and/or cooling system of a structure.

Title insurance. An insurance policy that protects ownership of a piece of real estate. Also called "clear title."

Trim out. The stage of construction when final trim-items are installed by each trade — toilets, moldings, light fixtures, etc.

Vapor barrier. A thin membrane impervious to moisture.

Wraparound. A new loan which includes an old low-interest loan, without having paid-off the old loan.

Additional Reading

So You Want to Fix Up an Old House, by Peter Hotton. Boston: Little, Brown, 1979.

This Old House. Alexandria, Virginia: Time-Life Books, 1980.

This Old House: Restoring, Rehabilitating & Renovation, by Bob Vila and Jane Davison. Boston: Little, Brown, 1980.

Index

Page references in *italics* indicate illustrations;
bold indicates charts and tables.

Other Storey Titles You Will Enjoy

Be Your Own House Contractor: Save 25% without Lifting a Hammer, by Carl Heldmann. In this book, learn trade secrets on buying land, making estimates, getting loans, picking subcontractors, and buying materials and supplies. 144 pages. Paperback. ISBN 0-88266-266-X.

Be Your Own Home Decorator: Creating the look you love without spending a fortune, by Pauline B. Guntlow. Presented with an infectious can-do attitude and clear step-by-step instructions, this useful book explains how to customize kitchens, bedrooms, living rooms, and baths. 144 pages. Paperback. ISBN 0-88266-945-1.

Reviving Old Houses: Over 500 Low-Cost Tips & Techniques, by Alan Dan Orme. This book offers practical advice on roofs, walls, masonry, glazing, insulation, plumbing, doors, stairs, floors, exteriors, and more. 180 pages. Paperback. ISBN 0-88266-563-4.

Small House Designs: Elegant, Architect-Designed Homes, 33 Award-Winning Plans, 1,250 Square Feet or Less, edited by Kenneth R. Tremblay, Jr. and Lawrence Von Bamford. This book offers cutting-edge designs produced by a worldwide array of architects, designers, and architecture students. 208 pages. Paperback, ISBN 0-88266-966-4; Hardcover, ISBN 0-88266-854-4.

Low-Cost Pole Building Construction: The Complete How-To Book, by Ralph Wolfe. This book includes building plans, drawings and materials lists, examples of already built homes, and detailed descriptions of pole building construction. 192 pages. Paperback. ISBN 0-88266-170-1.

Build a Classic Timber-Framed House: Planning & Design, Traditional Materials, Affordable Methods, by Jack A. Sobon. Using actual plans, this book shows how to build a classic hall-and-parlor home. 208 pages. Paperback. ISBN 0-88266-841-2.

Build Your Own Low-Cost Log Home, by Roger Hard. Features line drawings, photos, tables, and charts. It explains how to build a log home both using your own logs and with a kit. 208 pages. Paperback. ISBN 0-88266-399-2.